Series/Number 07-108

BASIC MATH FOR SOCIAL SCIENTISTS
Concepts

TIMOTHY M. HAGLE
The University of Iowa

SAGE PUBLICATIONS
International Educational and Professional Publisher
Thousand Oaks   London   New Delhi

For information address:

SAGE Publications, Inc.
2455 Teller Road
Thousand Oaks, California 91320
E-mail: order@sagepub.com

SAGE Publications Ltd.
6 Bonhill Street
London EC2A 4PU
United Kingdom

SAGE Publications India Pvt. Ltd.
M-32 Market
Greater Kailash I
New Delhi 110 048 India

Printed in the United States of America

Library of Congress Cataloging-in-Publication Data

Hagle, Timothy M.
Basic math for social scientists: concepts / Timothy M. Hagle.
p. cm. — (Sage university papers series. Quantitative applications in the social sciences; no. 07-108)
Includes bibliographical references.
ISBN 0-8039-5875-7 (pbk.: alk. paper)
1. Mathematics. 2. Social sciences—Mathematics. I. Title. II. Series.
QA37.2.H227 1995
515′.14—dc20 95-9081

95 96 97 98 99 10 9 8 7 6 5 4 3 2 1

Sage Project Editor: Susan McElroy

When citing a university paper, please use the proper form. Remember to cite the current Sage University Paper series title and include the paper number. One of the following formats can be adapted (depending on the style manual used):

(1) HAGLE, T. M. (1995) *Basic Math for Social Scientists: Concepts.* Sage University Paper series on Quantitative Applications in the Social Sciences, 07-108. Thousand Oaks, CA: Sage.

OR

(2) Hagle, T. M. (1995). *Basic math for social scientists: Concepts* (Sage University Paper series on Quantitative Applications in the Social Sciences, series no. 07-108). Thousand Oaks, CA: Sage.

CONTENTS

Series Editor's Introduction v

Math Symbols and Expressions vii

Greek Letters vii

Preface viii

1. Introduction **1**
- 1.1. Algebra Review 1
  - 1.1.1. Exponents 1
  - 1.1.2. Logarithms 2
  - 1.1.3. Relations and Functions 4
- 1.2. Sets 13
- 1.3. Permutations and Combinations 17

2. Limits and Continuity **22**
- 2.1. The Concept of a Limit 22
- 2.2. Infinite Limits and Limits at Infinity 25
- 2.3. Properties of Finite Limits 27
- 2.4. Continuity 30

3. Differential Calculus **31**
- 3.1. Tangents to Curves 33
- 3.2. Differentiation Rules 36
- 3.3. Extrema of Functions 42

4. Multivariate Functions and Partial Derivatives **47**
- 4.1. Partial Derivatives 47
- 4.2. Extrema of Multivariate Functions 50
  - 4.2.1. The Method of Least Squares 54
  - 4.2.2. Constrained Optimization (Method of Lagrange Multipliers) 56

5. Integral Calculus **58**
- 5.1. Integration Rules 59

5.2. The Theory of Integration and Definite Integrals 65

6. Matrix Algebra 71

6.1. Matrices 71

6.1.1. Matrix Rules 72

6.1.2. Representing a System of Equations in Matrix Form 76

6.2. Determinants 76

6.3. Inverses of Matrices 83

6.4. Cramer's Rule 86

6.5. Eigenvalues and Eigenvectors 89

6.6. Multivariate Extrema and Matrix Algebra 91

About the Author 96

SERIES EDITOR'S INTRODUCTION

As quantitative social scientists, we use statistics to describe, relate, or explain observed human variation. But statistics is a branch of mathematics, and to be comprehended fully its mathematical roots must be grasped. In this monograph, Dr. Hagle lays bear the basic math underlying the leading statistical procedures of social and behavioral research. Coverage is exhaustive, systematic, and lucid. His first chapter, an algebra review, refreshes us on exponents, logarithms, relations and functions, sets, and permutations and combinations. Upon development of the mathematical principle, it is immediately linked to the relevant data-analytic technique. Functions are regarded as a way to model relationships among variables. In particular, of course, is the linear function that structures regression analysis. There are many useful side remarks along the way, such as explanation of important notational conventions, for example, which letters usually represent functions and which represent variables.

To understand the "Least Squares Principle" for determining the coefficients of a regression equation (ordinary least squares—OLS), one needs a working knowledge of calculus fundamentals, explicated in the second and third chapters. The intuition for the central idea of a limit comes from his well-chosen example of trying to calculate Olympic runner Carl Lewis's speed at the halfway (50-meter) point. Averaging over ever-smaller measurement intervals around the 50-meter mark, we approach the value of the function, the runner's instantaneous speed. The third chapter has an explication of the technique of differentiation, which allows actual calculation of that instantaneous speed, or rate of change. Differentiation is widely used to solve optimization problems. Dr. Hagle conveniently summarizes the rules of differentiation in ten points, which are then followed by eight examples.

In the fourth chapter, functions with more than one independent variable are treated. Here, in order to decide the rate of change at multiple points, partial derivatives are taken. In partialling, all variables, save the one of interest, are held constant. Note that this resembles the language of regression analysis, a tie that Hagle makes explicit, observing that least squares is a minimization problem. In Chapter 5, the other basic operation of calculus—integration (the inverse of differentiation)—is presented. Integration enables us to establish the area under a probability curve.

The last chapter concentrates on matrix algebra, which we may apply to data arrayed in squares or rectangles. For example, with a 2×2 matrix, each column may stand for a variable, each row an observation, and each numerical entry the score on a particular variable and a particular observation. Certain statistical conditions, such as the meeting of the rank condition for identification in simultaneous equation systems, can only be tested using matrix algebra. In general, matrix algebra offers an efficient way of solving elaborate equation systems.

Our central statistical techniques in the various social science disciplines have a "pure math" core—algebra, calculus, matrix algebra. To do serious quantitative work, this mathematics needs to be mastered. The rigorous, clear exposition by Dr. Hagle provides an opportunity for such mastery. Indeed, the monograph would seem ideal to assign first semester graduate students, together with its companion volume, *Basic Math for Social Scientists: Problem and Solutions,* Vol. 109. (This second volume provides useful exercises on the concepts unfolded in the first volume.) Either book can be followed by students with no prior work in calculus or matrix algebra. Further, they provide a nice review for seasoned professors, who may have been away from the material for some time. Last, but by no means least, the books offer a comprehensive, low cost alternative to the texts commonly assigned for teaching these subjects.

—*Michael S. Lewis-Beck*
Series Editor

MATH SYMBOLS AND EXPRESSIONS

Symbol	Meaning
$\exists$	There exists
$\forall$	For all, for any
$\therefore$	Therefore
$\Rightarrow$	Implies
$\ni$	Such that
$\mid$	Such that
$\rightarrow$	Goes to, approaches
Δ	Delta, the change in
iff	if and only if
$\in$	Is an element of
$\notin$	Is not an element of
$\approx$	Nearly equal to
$\cong$	Equals approximately
$\neq$	Not equal to
$\times$	multiplied by
[ ]	Matrix, closed interval
$\mid \; \mid$	Absolute value, determinant
( )	Parentheses, open interval, point, matrix, combinations, etc.
{ }	Set
$\sum$	Summation
$\prod$	Product
∂	Partial differential
$\int$	Integral sign
!	Factorial sign
ln	Natural logarithm
$\log_b$	Logarithm to the base b
e	Base (≈ 2.718) of natural logarithms
π	Pi, ≈ 3.1416
μ	Mu, mean
σ	Sigma, standard deviation
σ^2	Sigma squared, variance
$\Re$	Real numbers
∞	Infinity
$\varnothing$	The empty set
$\cup$	Union sign
$\cap$	Intersection sign
$\subseteq$	Is a subset of
$\subset$	Is a proper subset of
lim	Limit of a function
$\circ$	Composition of two functions (small circle)

GREEK LETTERS

Α α Alpha	Ι ι Iota	Ρ ρ Rho
Β β Beta	Κ κ Kappa	Σ σ Sigma
Γ γ Gamma	Λ λ Lambda	Τ τ Tau
Δ δ Delta	Μ μ Mu	Υ υ Upsilon
Ε ε Epsilon	Ν ν Nu	Φ φ Phi
Ζ ζ Zeta	Ξ ξ Xi	Χ χ Chi
Η η Eta	Ο ο Omicron	Ψ ψ Psi
Θ θ Theta	Π π Pi	Ω ω Omega

PREFACE

The purpose of this monograph is to give students an introduction (or refresher) to many of the mathematical concepts and techniques that underlie quantitative analysis in the social sciences, not the least of which is regression analysis. My approach to the material is informal. There will be many definitions, equations, and examples as well as alternative notation and supplemental information, but few proofs. Those wishing a deeper understanding of the material can turn to more complete treatments of these topics contained in mathematics textbooks that are available in any college bookstore.

The monograph comprises six chapters. Chapter 1 contains an introductory algebra review and material on sets and combinations. Chapter 2 contains a discussion of limits and continuity. Chapter 3 begins the calculus material with a presentation of differential calculus. Chapter 4 continues the calculus material with a discussion of multivariate functions, partial derivatives, and multivariate extrema. Chapter 5 concludes the calculus material with a presentation of integral calculus. Chapter 6 contains a discussion of matrix algebra. A list of common math symbols and expressions and a list of upper and lowercase Greek letters are included on p. vii.

The reader need not have had a previous course in calculus or matrix algebra to understand the material in this monograph. I do, however, assume the reader is familiar with basic algebra concepts. The algebra review in Chapter 1 is primarily for those who may not have used their math skills recently.

BASIC MATH FOR SOCIAL SCIENTISTS
Concepts

TIMOTHY M. HAGLE
University of Iowa

1. INTRODUCTION

This chapter contains a review of some topics with which one should be familiar before proceeding to the calculus material. The chapter begins with a brief algebra review followed by sections discussing sets and combinations.

1.1 Algebra Review

1.1.1 Exponents

An exponent is a number placed in a superscript position immediately following another number or variable to indicate repeated multiplication. Thus $2 \times 2 \times 2 = 2^3$. More generally, $a \times a \times a = a^3$ and $a \times a \times \ldots \times a$ (n times) $= a^n$. Here are 12 rules for using exponents:

1. $a = a^1$; if the exponent is 1, it is usually assumed and not written, e.g., $2^1 = 2$
2. $a^m \times a^n = a^{m+n}$; e.g., $2^2 \times 2^3 = 2^{2+3} = 2^5 = 32 = 4 \times 8 = 2^2 \times 2^3$
3. $(a^m)^n = a^{m \times n}$; e.g., $(2^2)^3 = 2^6 = 64 = (4)^3 = (2^2)^3$
4. $(a \times b)^n = a^n \times b^n$; e.g., $(2 \times 3)^2 = 2^2 \times 3^2 = 4 \times 9 = 36 = 6^2 = (2 \times 3)^2$
5. $\left(\frac{a}{b}\right)^n = \frac{a^n}{b^n}$, for $b \neq 0$; e.g., $\left(\frac{4}{2}\right)^2 = \frac{4^2}{2^2} = \frac{16}{4} = 4 = 2^2 = \left(\frac{4}{2}\right)^2$

6. $a^{-n} = \frac{1}{a^n}$; e.g., $2^{-3} = \frac{1}{2^3} = \frac{1}{8}$

7. $\frac{a^m}{a^n} = a^{m-n} = \frac{1}{a^{n-m}}$; e.g., $\frac{2^2}{2^3} = 2^{2-3} = 2^{-1} = \frac{1}{2^1}$

8. $a^{1/2} = \sqrt{a}$, because (from rule 2) $a^{1/2} \times a^{1/2} = a^{1/2 + 1/2} = a^1 = a$

9. $a^{1/n} = \sqrt[n]{a}$; e.g., $8^{1/3} = \sqrt[3]{8} = 2$

10. $a^{m/n} = (a^{1/n})^m = (a^m)^{1/n} = \sqrt[n]{a^m} = (\sqrt[n]{a})^m$; e.g., $4^{3/2} = (4^{1/2})^3 = (4^3)^{1/2} = \sqrt[2]{4^3} = (\sqrt[2]{4})^3 = 8$ (considering only the positive root)

11. $a^0 = 1$, because $a^0 = a^{n-n} = (a^n)(a^{-n}) = \frac{a^n}{a^n} = 1$; e.g., $2^0 = 1$

12. 0^0 is undefined (though it may be specially defined for a given context)

Numbers 8, 9, and 10 use the radical sign ($\sqrt{\ }$) to show how, for example, the square root of a number is the same as that number raised to the one-half power. The use of radicals can become messy, as you can see from number 10 above. You can usually make calculations easier by converting radicals to exponents.

1.1.2 Logarithms

For a positive number *n*, the logarithm of *n* is the power to which some number *b* (the base of the logarithm) must be raised to yield *n*. Thus if $b^x = n$, then $\log_b n = x$.

This simply means that if *x* is the logarithm of *n*, then you must raise the base *b* to the *x* power to get *n*. *Common logarithms* use 10 as the base.

The common logarithm (or just "log") of 100 is 2 because $10^2 = 100$. We can also write this as $\log_{10} 100 = 2$. Traditionally, $\log_{10} n$ is written as $\log n$. Another widely used base is *e* (≈ 2.718). Logarithms to the base *e* are called *natural logs* and are usually written $\ln n$ (rather than $\log_e n$). Because economists and social scientists are more likely to use the natural log, they often use *log*, rather than *ln*, to represent the natural log in their equations. One can easily adapt to either usage if the authors indicate how they are using the symbols. Many times, however, they do not. If a formula uses *log* but does not produce the proper results using the common log, try using the natural log.

Here are three basic rules for using logarithms:

1. $\log(a \times b) = \log a + \log b$; e.g., $\log(10 \times 100) = \log 10 + \log 100 = 1 + 2 = 3 = \log 1000$
2. $\log(a/b) = \log a - \log b$; e.g., $\log(1000/10) = \log 1000 - \log 10 = 3 - 1 = 2 = \log 100$
3. $\log(a^n) = n \log a$; e.g., $\log(10^3) = 3 \log 10 = 3 \times 1 = 3 = \log 1000$

Here are two examples that are not quite so neat:

1. $\left(\frac{1}{3}\right)\log 64 = \log\left(64^{1/3}\right) = \log 4 \approx .6021$
2. $2\ln 3 - \ln 7 = \ln(3^2) - \ln 7 = \ln\left(\frac{3^2}{7}\right) = \ln\left(\frac{9}{7}\right) \approx \ln 1.2857 \approx .2513$

Here are two rules that show the relationship between exponents and logarithms (which are inverse functions of each other):

1. $\log_a a^x = x$; $\ln e^x = x$
2. $a^{\log_a x} = x$; $e^{\ln x} = x$

With a good calculator (strongly recommended) you can get the answer without manipulating the equation, but manipulating equations and formulas will help you to understand the rules and relationships of a concept. In later chapters you will see that the ability to manipulate an equation to change it from an unfamiliar form to a familiar one is a valuable problem-solving technique.

Logarithms are used in several ways in quantitative analysis. Two of the more important ones are worth mentioning here. First, many social scientists use some form of regression analysis to examine their data. In essence, regression analysis tries to find the line that best fits the data. Of course, not all data fit a linear model. By using logarithms it may be possible to transform a nonlinear model into a linear form so the researcher can make use of well-developed regression techniques and diagnostics. For example, consider the theoretical model $y = \alpha x_1^{\beta_1} x_2^{\beta_2}$, where y is the dependent variable, the x_i are independent variables, the β_i are unknown powers of the independent variables, and α is a constant. We can transform the right side of the equation to a linear form by taking the log of both sides of the equation. Thus

$$\log y = \log(\alpha x_1^{\beta_1} x_2^{\beta_2}) = \log\alpha + \beta_1 \log x_1 + \beta_2 \log x_2$$

The researcher can convert the independent variable data, use regression techniques to estimate α and the β_i, and later convert the data back to their original form.

Logarithms are also useful when the values for a particular independent variable are very large, particularly when the values of the other independent variables are small by comparison. Taking the log of large values to "shrink" them to a more manageable size will keep them from unduly influencing the results of the analysis.

1.1.3 Relations and Functions

We first consider functions and relations in two dimensions. To graph these functions and relations we normally use the Cartesian coordinate system (after French mathematician René Descartes). The Cartesian coordinate system is sometimes referred to as the Cartesian plane and more often as the *xy*-plane. You may occasionally see the Cartesian plane designated as $\Re^2$, where $\Re$ stands for "real" numbers and the superscript indicates the number of dimensions. Points in the plane are designated by an ordered pair (x, y), where x indicates the distance of the point from the origin (the intersection of the x and y axes) along the (horizontal) x-axis. Similarly, y represents the distance of the point from the origin along the (vertical) y-axis.

A *relation*, which we designate as R, is an association between two or more objects. We shall begin by considering relations in the Cartesian plane involving values of x and y. Thus R is a subset of $\Re^2$. The *domain* of the relation is the set of all possible x values of the relation. The *range* of the relation is the set of all possible y values of the relation. The domain and range can be formally written as

$$D_{\mathrm{R}} = \{x \mid \exists\, y \ni (x, y) \in \mathrm{R}\} \quad \text{and} \quad R_{\mathrm{R}} = \{y \mid \exists\, x \ni (x, y) \in \mathrm{R}\}$$

The domain statement is read as: The domain of the relation R consists of the set x such that there exists y such that the ordered pair (x, y) is an element of the relation R. The range statement is read in a similar fashion.

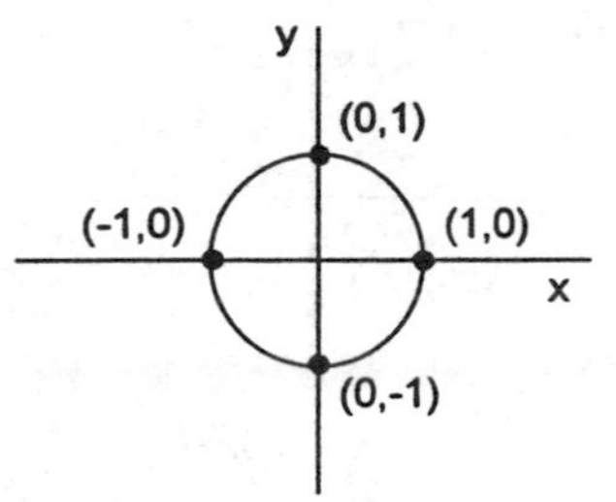

Figure 1.1. The Unit Circle: $x^2 + y^2 = 1$

Example: Consider the relation $x^2 + y^2 = 1$ whose domain and range are as follows:

$$D_R = \{x \mid -1 \leq x \leq 1\} \quad \text{and} \quad R_R = \{y \mid -1 \leq y \leq 1\}$$

We can see from Figure 1.1 that the graph of this relation is a circle of radius 1 that is centered at the origin. This relation is often referred to as the unit circle.

A *function* is a special type of relation such that no two members of the relation have the same first coordinate (x value). Functions can be denoted in several ways. For example, $f: \Re \rightarrow \Re$ tells us that the function goes from the real numbers to the real numbers. This relationship can also be expressed as $\Re \xrightarrow{f} \Re$. The most familiar notation for functions is $y = f(x)$, which tells us that a function, comprising one or more operations or transformations, is performed upon the values of x (the domain) to obtain the values of y (the range).

Examples: Which of the following relations are functions? What are their domains and ranges? Graph the relations:

(a) $y - x^2 = 0 \quad \Rightarrow \quad y = x^2$

(b) $xy = 1 \quad \Rightarrow \quad y = \dfrac{1}{x}$

(c) $x^2 - y^2 = 1 \quad \Rightarrow \quad y = \pm\sqrt{x^2 - 1}$

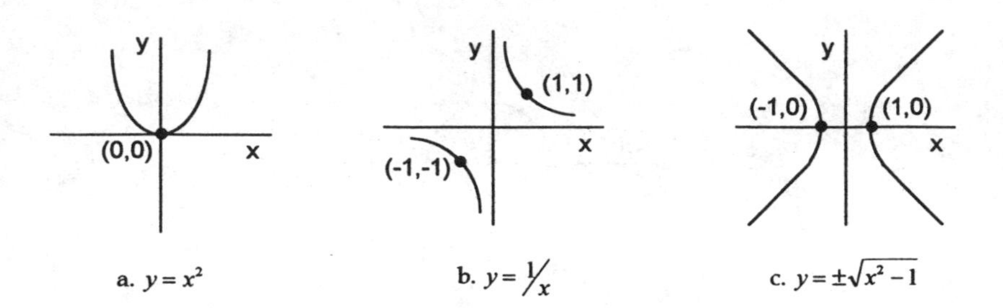

Figure 1.2. Graphs of Three Relations

It is easier to work with relations in $y = f(x)$ form, so I rewrote each example in that form. Next construct a list of some x and y values for each relation.

(a) $y = x^2$		(b) $y = \frac{1}{x}$		(c) $y = \pm\sqrt{x^2 - 1}$	
x	y	x	y	x	y
−2	4	−2	−½	−2	$\pm\sqrt{3}$
−1	1	−1	−1	−1	0
0	0	0	undefined	0	not in reals
1	1	1	1	1	0
2	4	2	½	2	$\pm\sqrt{3}$
$D = \Re$		$D = \{x \mid x \neq 0\}$		$D = \{x \mid x \leq -1 \text{ or } x \geq 1\}$	
$R = \{y \mid y \geq 0\}$		$R = \{y \mid y \neq 0\}$		$R = \Re$	

We can plot these x and y values to get a general idea of the shape of the graph. The graphs of these examples are shown in Figure 1.2. If we could not tell from the equation for the first example, its graph shows that x may take on all real numbers (the domain) whereas y is limited to real numbers greater than or equal to zero (the range). The domain and range of each relation are listed at the bottom of each set of values.

From the first example's equation, we can see that it is not possible for the same x value to yield two different y values. (We do have two x values yielding the same y value in this example, but that does not matter in determining whether a relation is a function.) If the equation is complex, we may not be able to tell just from looking whether it is a function. If so, we can examine a graph of the relation. If you can draw a vertical line anywhere on the graph so that the graph crosses the line in at least two points, then the relation is *not* a function. Using this method on the first example again shows it to be a function.

The five pairs of values for the other two examples will not provide a very complete picture of the relation. One should, of course, choose as many x values as is necessary to obtain a complete picture of the graph. From the graphs in Figures 1.2b and 1.2c, we can see that the second example is a function and the third is not. In the third example, there is a positive and a negative square root of $x^2 - 1$. Because the same value for x generates two y values, the relation is not a function. We can, however, consider the relation to be a function if we ignore the negative square root. We sometimes ignore the negative root when a negative value for y does not make sense in the context of the problem (e.g., if y represents time).

Functions are valuable to social scientists because most relationships can be modeled in the form of a function. In addition, functions can be manipulated much like other algebraic expressions. For example:

1. $f \pm g : x = f(x) \pm g(x)$
2. $a \times f : x = a[f(x)]$ for $a \in \Re$
3. $f \times g : x = f(x) \times g(x)$
4. $f^a : x = [f(x)]^a$ for $a \in \Re$
5. $f/g : x = f(x)/g(x)$, provided $g(x) \neq 0$
6. $g \circ f : x = g[f(x)]$

The first five operations are addition or subtraction of two functions, multiplication of a function by a constant, multiplication of two functions, division of two functions, and raising a function to a power. The sixth operation is the *composition* of two functions. Given two functions $f:\Re \to \Re$ and $g:\Re \to \Re$, we define the composition of f and g to be the function from $\Re$ to $\Re$ as follows: $\forall\, x \in \Re,\ g \circ f : x = g[f(x)]$. The composition $g \circ f : x$ is read as "g of f of x" or "g composed with f." We might represent the composition somewhat more visually as $\underbrace{\Re \xrightarrow{f} \Re \xrightarrow{g} \Re}_{g \circ f}$.

Examples: If $f(x) = x^2 + 3x$ and $g(x) = 2x + 5$,

1. $f + g : x = \overbrace{x^2 + 3x}^{f(x)} + \overbrace{2x + 5}^{g(x)} = x^2 + 5x + 5$
2. $3 \times f : x = 3(x^2 + 3x) = 3x^2 + 9x$
3. $f \times g : x = (x^2 + 3x)(2x + 5) = 2x^3 + 6x^2 + 5x^2 + 15x$
   $= 2x^3 + 11x^2 + 15x$
4. $f^2 : x = (x^2 + 3x)^2 = (x^2 + 3x)(x^2 + 3x) = x^4 + 6x^3 + 9x^2$
5. $g \circ f : x = 2(x^2 + 3x) + 5 = 2x^2 + 6x + 5$
6. $f \circ g : x = (2x + 5)^2 + 3(2x + 5) = 4x^2 + 20x + 25 + 6x + 15$
   $= 4x^2 + 26x + 40$

There are two things you should notice about the last two examples. First, to calculate the composition we substitute the first function into the second function wherever we see x in the second function. Thus to determine $g \circ f : x$ we substituted (or "plugged in") $f(x) = x^2 + 3x$ for every x in $g(x) = 2x + 5$ and then simplified by doing the multiplication. Second, here, and in general, $g \circ f : x \neq f \circ g : x$.

One of the most important functions (particularly for those using regression analysis) is the *linear function*. As the name suggests, the graph of a linear function is a straight line. Algebraically, linear functions take the form $y = mx + b$, where y is a function of x, m is the slope of the line, and b is the y-intercept (i.e., where the line crosses the y-axis). You may also see linear functions written as $y = b_0 + b_1x_1$. In this form, b_0 is the constant and b_1 is the slope. The subscript on the independent variable, x, does not serve much of a purpose when there is only one independent variable. Subscripts become more important when there are more independent

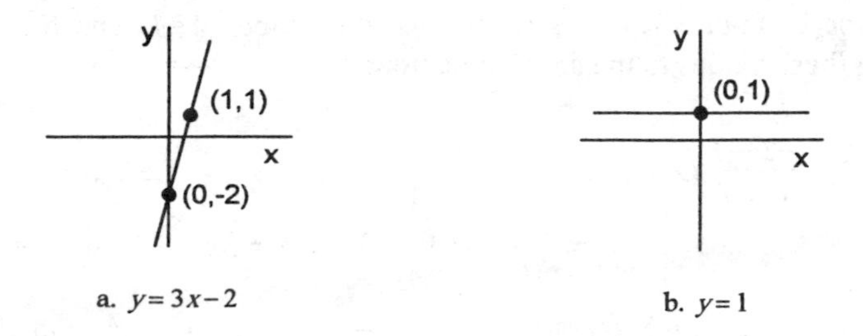

a. $y = 3x - 2$

b. $y = 1$

Figure 1.3. Graphs of Two Linear Functions

variables and we must distinguish between them without having to choose a different letter for each. For example,

$$y = b_0 + b_1x_1 + b_2x_2 + b_3x_3 = \sum_{i=0}^{3} b_ix_i$$

has three independent variables. This function is linear because each of the independent variables is raised only to the first power, but it is in four dimensions (one for each of the three independent variables and one for the dependent variable). We can simplify writing the function by writing it as a summation. Given the way summations are read, the first term is really b_0x_0, but we usually assume that $x_0 = 1$ and do not write the x_0. Alternatively, the function can be written as $b_0 + \sum_{i=1}^{3} b_ix_i$, where i begins with 1 rather than 0.

A special form of linear function is the *constant function*. In a constant function, the slope of the line is zero. Thus for $y = 0x + b$ the first term is always equal to 0 and $y = b$ regardless of the value of x (which is why it is called a constant function). The graph of a constant function is a horizontal line that crosses the y-axis at b. The graphs of two linear functions (the second a constant function) are presented in Figure 1.3.

Returning to the simple linear function, consider the function $y = 3x - 2$. The graph of this function appears in Figure 1.3a. Given the coordinates of two points, you can determine the equation of the line that passes through them. For points (x_1, y_1) and (x_2, y_2), the equation of the line is

$$y - y_1 = \frac{y_2 - y_1}{x_2 - x_1}(x - x_1)$$

The line in Figure 1.3a passes through the points (1, 1) and (0, –2). Entering these values in the above equation:

$$y - 1 = \frac{-2 - 1}{0 - 1}(x - 1) \Rightarrow y - 1 = \frac{-3}{-1}(x - 1) \Rightarrow y - 1 = 3(x - 1)$$

$$\Rightarrow y - 1 = 3x - 3 \Rightarrow y = 3x - 2$$

Of course, the last expression, $y = 3x - 2$, is the original equation. Although this example is rather trivial, the principle is quite important to regression analysis. Regression analysis *assumes* a linear relationship between the dependent variable (y) and the independent variable(s) (x). The major thrust of regression analysis is to find the line that is the best fit for several data points. As we will see in Chapter 4, the form of regression analysis known as "Ordinary Least Squares" (OLS) uses a technique quite similar to the one shown above.

Example: Find the equation of the line that passes through the points (2, –1) and (–1, 1). Using the formula from above:

$$y - (-1) = \frac{1 - (-1)}{-1 - 2}(x - 2) \Rightarrow y + 1 = -\tfrac{2}{3}(x - 2) \Rightarrow y + 1 = -\tfrac{2}{3}x + \tfrac{4}{3}$$

$$\Rightarrow y = -\tfrac{2}{3}x + \tfrac{1}{3}$$

From this equation you can tell that the line crosses the y-axis at (0, ⅓). The negative slope tells us that the line moves downward as one moves from left to right along the x-axis. (Graph the two points to verify this answer.)

We can use our ability to manipulate functions when we want to "solve for a system of linear equations." This means that we have more than one line and want to know whether they cross, and, if so, at what point. Generally, we can solve for a system of linear equations provided that we have at least as many equations as unknowns (independent variables).

Example: Find the point where $y = 3x - 2$ and $y = -\tfrac{2}{3}x - 2$ intersect. Because both equations are equal to y, substitute the right-hand portion of the first equation for y in the second. This yields $3x - 2 = -\tfrac{2}{3}x - 2$. Now solve for x as follows:

$$3x - 2 = -\tfrac{2}{3}x - 2 \Rightarrow 9x - 6 = -2x - 6 \Rightarrow 11x = 0 \Rightarrow x = 0$$

Entering this value into one of the equations tells us that $y = -2$, and the point where the two lines intersect is (0, –2).

Example: Find the point where $y = 2x + 5$ and $y = -x - 3$ intersect. Again, set the two equations equal to each other and solve for x:

$$2x + 5 = -x - 3 \Rightarrow 3x = -8 \Rightarrow x = -\tfrac{8}{3}$$

$$y = -(-\tfrac{8}{3}) - 3 = \tfrac{8}{3} - \tfrac{9}{3} = -\tfrac{1}{3}$$

Thus the two lines intersect at $(-\tfrac{8}{3}, -\tfrac{1}{3})$. Verify this result by graphing the two lines.

You may have noticed in the first example that the y-intercept for both equations was –2; that is, both lines cross the y-axis at (0, –2). Because lines can intersect at only one point, we should have known that (0, –2) would be the solution. You could have manipulated the equations in other ways to reach the same solution. If you have more unknowns and more equations, the manipulation will be more difficult. Examples with more unknowns appear in later chapters. In general, your strategy should be to isolate one of the variables, find its value, and use it to find the values for the other variables.

The final type of function with which we will be concerned is the *polynomial function*. The difference from the linear function is that the powers of the variables need only be nonnegative integers. The general form of a polynomial function in one variable is

$$a_0 + a_1x + a_2x^2 + a_3x^3 + \ldots + a_nx^n = \sum_{i=0}^{n} a_ix^i = a_0 + \sum_{i=1}^{n} a_ix^i$$

As with the linear function, because the power of x in the first term is 0, x^0 may or may not be written and may or may not be inside the summation depending on whether the summation begins with 0 or 1.

There will be times when we will want to know the values of x for which a polynomial is equal to 0, that is, when $f(x) = 0$. To find such values of x, we factor the polynomial into linear factors and set each equal to 0. There is no trick in learning to factor polynomials, it just takes practice. When you have done it many times, you begin to recognize patterns that allow you to do the factoring more quickly. For practice, begin by multiplying some factors together.

Example: $(3x - 2)(2x + 1) = 6x^2 + 3x - 4x - 2 = 6x^2 - x - 2$

Example: $(ax + b)(cx + d) = acx^2 + (bc + ad)x + bd$

If we want to know when $6x^2 - x - 2 = 0$ we need only determine when $3x - 2 = 0$ or $2x + 1 = 0$. If either factor equals 0, then the polynomial of which it is a part also equals 0. Thus the polynomial of the first example is equal to 0 when

$$3x - 2 = 0 \Rightarrow 3x = 2 \Rightarrow x = 2/3$$
$$2x + 1 = 0 \Rightarrow 2x = -1 \Rightarrow x = -1/2$$

Check these values in the original equation:

$$6(2/3)^2 - (2/3) - 2 = 6(4/9) - 8/3 = 24/9 - 24/9 = 0$$
$$6(-1/2)^2 - (-1/2) - 2 = 6(1/4) + 1/2 - 2 = 6/4 - 3/2 = 6/4 - 6/4 = 0$$

Examples: Factor to solve the following:

1. $0 = x^2 + 7x + 10 = (x + 2)(x + 5) \Rightarrow x = -2, -5$
2. $0 = 3x^2 - 8x - 3 = (x - 3)(3x + 1) \Rightarrow x = 3, -1/3$

Not all polynomials will be so easy to factor. If we have a polynomial of the form $0 = ax^2 + bx + c$, then we can use the *quadratic formula* to solve for x:

$$x = \frac{-b \pm \sqrt{b^2 - 4ac}}{2a}$$

Using the second example from above:

$$x = \frac{-(-8) + \sqrt{(-8)^2 - 4(3)(-3)}}{2(3)} \qquad x = \frac{-(-8) - \sqrt{(-8)^2 - 4(3)(-3)}}{2(3)}$$

$$= \frac{8 + \sqrt{64 + 36}}{6} \qquad = \frac{8 - \sqrt{64 + 36}}{6}$$

$$= \frac{8 + \sqrt{100}}{6} \qquad = \frac{8 - \sqrt{100}}{6}$$

$$= \frac{8 + 10}{6} \qquad = \frac{8 - 10}{6}$$

$$= \frac{18}{6} \qquad = \frac{-2}{6}$$

$$= 3 \qquad = -1/3$$

Note: You may have noticed (or already be aware) that certain letters are repeatedly used in some contexts whereas other letters are used in other contexts. Although there may be variations across disciplines, the letters *a*, *b*, *c*, and *d* are usually used to represent constants; *f*, *g*, and *h* usually represent functions; *i*, *j*, and *k* are usually used as place holders or term identifiers as in summations; *m* and *n* usually represent numbers; *p* and *q* often represent probabilities; and *x*, *y*, *z*, and *w* usually represent variables. In addition, upper and lowercase Greek letters are often used to represent variables and quantities (see p. vii). Of course, these and other letters can be used however one wishes provided they are properly defined for the reader. Most authors will try to be consistent in using letters to represent quantities. Knowing this should help you in understanding the equations.

1.2 Sets

Some statisticians refer to a process of measurement or observation as an *experiment*. (This is a broader usage of the term than is generally used by social scientists.) Such an experiment can consist of something as simple as determining whether a switch is turned on or off, or as complex as determining who will win the next presidential election. The measurement or observation obtained from an experiment is called an *outcome*. We often represent the possible outcomes of an experiment as a set of points.

Examples: Consider the outcomes that can result from flipping a coin once. The coin can come up either a head or a tail. We can represent these possible outcomes on a straight line using two points as in Figure 1.4a. As

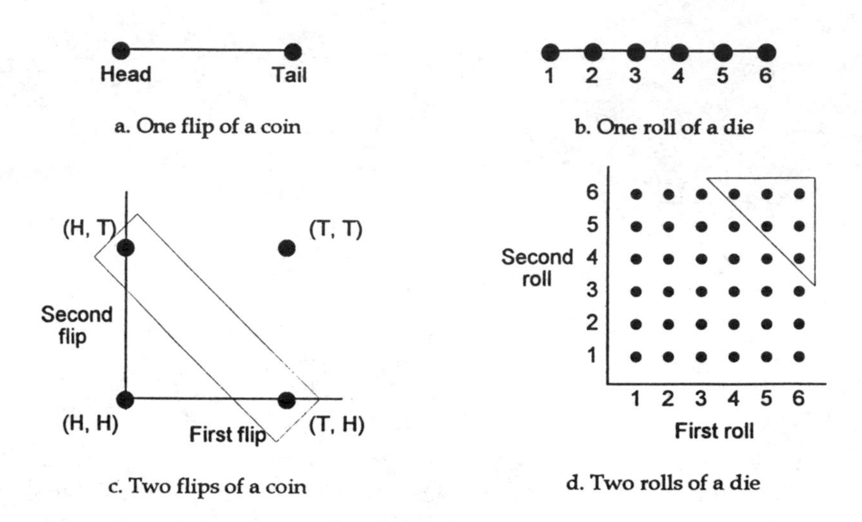

Figure 1.4. Representations of Outcomes and Sets

a second example, note that the outcomes from rolling a die can be represented by a line with six points as in Figure 1.4b. If we flip the coin twice, the number of heads occurring can be represented by three points: 0, 1, and 2. If we roll the die twice, the total number of dots appearing can be represented by eleven points consisting of 2 through 12. Suppose, however, that we also want to know the outcome for each trial, as opposed to just the total number of heads or dots. By keeping track of the outcome of each trial, it now takes four points to represent all the possible outcomes of two coin flips. We might get a head on the first flip and a tail on the second (H, T), or a tail on the first and a head on the second (T, H), or two heads (H, H), or two tails (T, T). This set of outcomes is represented graphically in Figure 1.4c. Figure 1.4d shows the 36 possible outcomes for two rolls of a die.

A *set* is loosely defined as a collection of points (or other objects). Points or objects belonging to a set are usually referred to as *elements* of the set. When the number of elements in a set is small they may all be listed in a set of braces ({ }). For example, the number of heads that may appear in two flips of a coin can be represented as {0, 1, 2}. If we want to distinguish between the first and second flip, we can represent the set as {(H, H), (H, T), (T, H), (T, T)}. Clearly, it would be too cumbersome to always list every element of a set. We can also designate a set by some rule that defines its elements. For example, we can define the set of all even integers as: $S = \{s \mid s = 2k, k = 1, 2, \ldots\}$.

If the set designates all possible outcomes, it is often called the *universe* (or universal set or sample space) and is denoted by U. A set U is considered *discrete* if it contains a finite number of elements or contains a "countable infinity" of elements (e.g., the set of whole positive integers). If U contains a continuum, such as all the points on a line, it is said to be *continuous*.

Note: Social scientists try to use universes whose elements are as basic as possible. More information is made available by using the most basic elements. For example, in the experiment of counting the number of heads that appear in two flips of a coin, the universe represented in Figure 1.4c is preferable to the universe represented by the line with three points because it contains more information, namely, the outcome of each trial. If we do not need the additional information of what occurred on each trial, it is easy to collapse the data to the other form. If we begin with the line with just three points, however, we cannot determine the outcome of each trial. (With just two trials you could probably remember and reconstruct the outcomes. This would be less likely if there were 100 trials, and probably impossible if you were not the one who originally collected the data.)

A portion of the elements in U is called a *subset* of U. In other words, the set A is called a subset of set B if and only if (iff) each element of A is also an element of B. More formally, $A \subseteq B$ iff $\forall\, x \in A, x \in B$. If the set A is a *proper subset* of U (i.e., $A \subset U$) then there are elements contained in U that are not contained in A (i.e., $A \subset B \Rightarrow \exists\, x \in B \ni x \notin A$). An *event* is a particular set of outcomes, usually a subset of all possible outcomes. For example, the area enclosed by the rectangle in Figure 1.4c contains the event of only one head appearing in the two trials. The area enclosed by the triangle in Figure 1.4d contains the event of the total of the two rolls of the die equaling 10 or more.

If set A belongs to a universe U, then A', called the *complement* of A, is the set of all the elements of U that are not elements of A; that is, if $A \subset U$, then $A' = \{x \mid x \in U \ni x \notin A\}$. For example, if U is the set of outcomes from rolling two dice, and A is the set of outcomes where the total is 10 or more, then A' is the set of outcomes where the total of the dice is nine or less (i.e., all the points not within the triangle in Figure 1.4d).

New sets can be formed by combining the elements of existing sets. The *intersection* of two sets A and B, denoted $A \cap B$ and read "the intersection of A and B," consists of the set of elements that belong to both A and B (i.e., $A \cap B = \{x \mid x \in A \text{ and } x \in B\}$). If two sets have no elements in common, they are said to be *disjoint*. For example, and by definition, A and A' are disjoint (i.e., $A \cap A' = \varnothing$, where $\varnothing$ means the *empty set*—a set containing no elements). If two sets are disjoint, the events associated with those sets are said to be *mutually exclusive*. (Whether events are mutually exclusive is important in examining the relationships between independent variables and between independent and dependent variables.) The *union* of two sets, denoted $A \cup B$ and read "the union of A and B," consists of all the elements that belong to either A or B (i.e., $A \cup B = \{x \mid x \in A \text{ or } x \in B\}$). (Elements that are in both A and B are not counted twice.)

Examples: Let U be the set of all American citizens.

1. If A is the set of all Americans who are registered to vote and B is the set of all Americans who are older than 65, then $A \cap B$ is the set of all Americans older than 65 who are registered to vote.
2. If A is as above and C is the set of Americans who are younger than 18, then the sets A and C are disjoint and $A \cap C = \varnothing$ (provided no one under 18 is illegally registered to vote).
3. If A and B are as above, then $A \cup B$ is the set of all Americans who are either registered to vote or over 65 years old.
4. The set $(A \cup B)'$ is the complement of $A \cup B$ and consists of all elements of U that are in neither A nor B. If A and B are as above, $(A \cup B)'$ consists of those Americans who are 65 or younger and not registered to vote.

Universes, subsets, and events are often depicted by means of *Venn diagrams*. Figure 1.5 contains four combinations of sets and subsets. The universe is enclosed by the box in these diagrams. Sometimes you will see Venn diagrams without such a box and must assume everything outside the represented sets constitutes the universe.

Set Rules: We consider sets to be mathematical objects that behave according to certain rules. Here are six rules for sets.

1. For each pair of sets A and B there exist unique sets $A \cup B$ and $A \cap B$ in the universe U: Closure.
2. $A \cup B = B \cup A$ and $A \cap B = B \cap A$: Commutative.

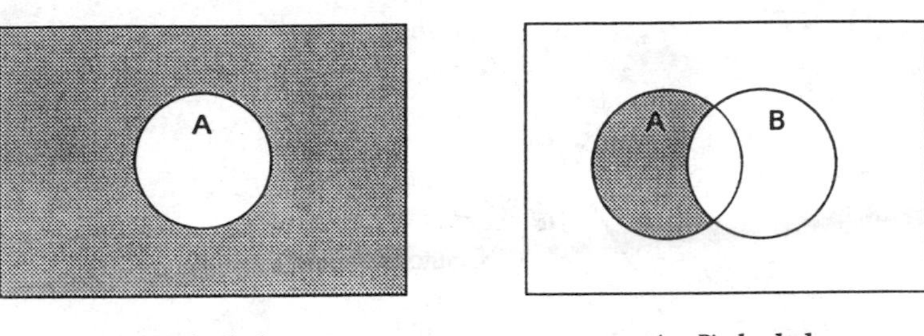

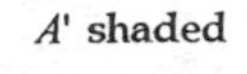

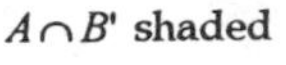

A' shaded

$A \cap B'$ shaded

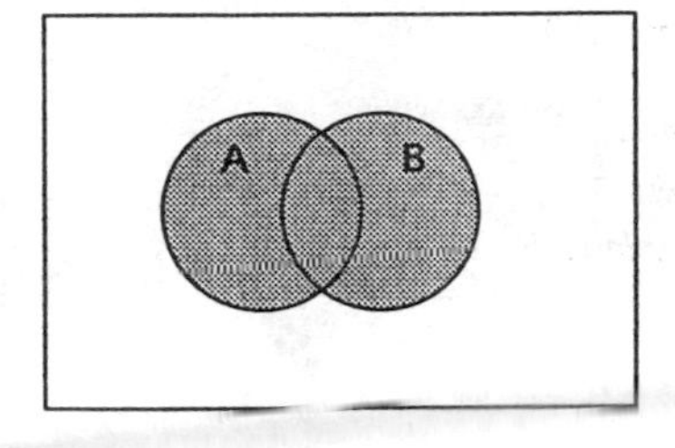

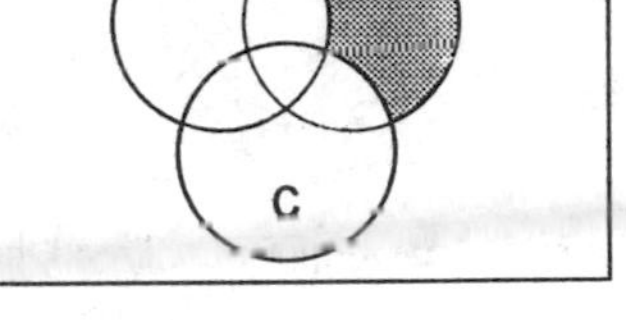

$A \cup B$ shaded

$B \cap (A \cup C)'$ shaded

Figure 1.5. Examples of Venn Diagrams

3. $(A \cup B) \cup C = A \cup (B \cup C)$ and $(A \cap B) \cap C = A \cap (B \cap C)$: Associative.
4. $A \cap (B \cup C) = (A \cap B) \cup (A \cap C)$ and $A \cup (B \cap C) = (A \cup B) \cap (A \cup C)$: Distributive.
5. $A \cap U = A$ for each set $A \subseteq U$, and there exists a unique set $\varnothing$ such that $A \cup \varnothing = A$ for each set A: Identity.
6. For each set $A \subseteq U$ there exists a unique set A' such that $A \cap A' = \varnothing$ and $A \cup A' = U$: Complementation.

1.3 Permutations and Combinations

Before we can determine the probability that a given event will occur, we must know what events are possible. In other words, we must be able to determine all possible outcomes for a particular situation. In addition to being able to describe or identify all possible outcomes, we must be able to determine how many possible outcomes exist.

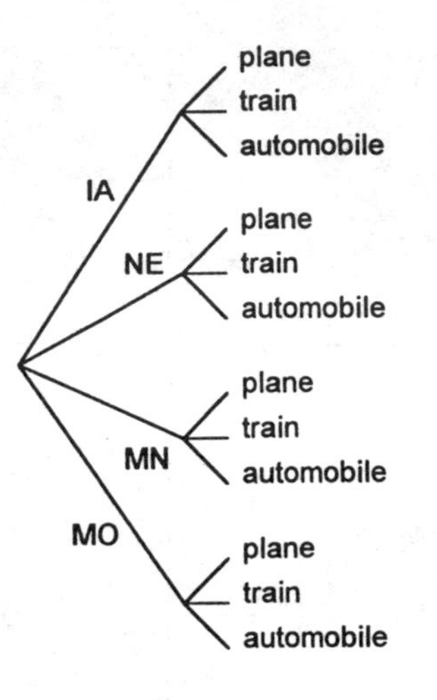

Figure 1.6. Example of a Tree Diagram

Tree diagrams are often used to list the possible outcomes. For example, Figure 1.6 shows the possible outcomes for a traveler who has a choice of four states and three modes of transportation. As drawn, the traveler's first decision is the destination and the second decision is the mode of transportation. There are 12 possible outcomes: going to Iowa by plane, train, or automobile; going to Nebraska by plane, train, or automobile; going to Minnesota by plane, train, or automobile; or going to Missouri by plane, train, or automobile. The four choices for a destination and three choices for a mode of transportation to each destination combine to yield $4 \times 3 = 12$ different outcomes.

If you have n elements in set A and m elements in set B, there are $n \times m$ different ways that you can select one element from set A and one from set B. More generally, if sets $A_1, A_2, \ldots, A_m$ have, respectively, $n_1, n_2, \ldots, n_m$ elements, then there are $n_1 \times n_2 \times \ldots \times n_m$ different ways of selecting one element from each set.

Example: If there are 12 girls and 9 boys in a choir, there are $12 \times 9 = 108$ different ways one girl and one boy can be selected for a duet.

Example: If a car dealership offers six models, each of which has eight color choices, and there are three different financing plans, then there are $6 \times 8 \times 3 = 144$ different ways to choose a car, color, and financing plan.

The problem in *permutations* usually involves the *order* in which several elements are selected from the same set. Consider the traveler from above. Suppose the traveler is a sales representative who must visit each of the states. Without worrying about the mode of transportation, how many different ways can the traveler visit each of the states? Obviously, there are four choices for the first state. After choosing the first state, only three choices remain for the second state, then two for the third state, and one for the last state. Thus there are $4 \times 3 \times 2 \times 1 = 24$ different orders in which the states can be visited on the sales trip. This example assumes *no replacement*, that is, that each state will be visited only once. In general, a particular selection of r objects chosen from a set of n objects is called a permutation. The number of permutations that can be formed by selecting r objects from a set of n objects is $n \times (n-1) \times (n-2) \times \ldots \times [n-(r-1)]$ and is denoted $P(n, r)$. In factorial notation, the formula is written as

$$P(n, r) = \frac{n!}{(n-r)!}$$

where n and r are positive integers and $r \leq n$. (*Note:* Recall that $n! = n \times (n-1) \times (n-2) \times \ldots \times 2 \times 1$ and, by definition, $0! = 1$.)

Example: Suppose candidates for a runoff election are placed on the ballot in the order of their initial vote total. How many ways can three candidates from a field of eight be placed on the ballot for a runoff election? Using the formula from above:

$$P(8, 3) = \frac{8!}{(8-3)!} = \frac{8!}{5!} = \frac{8 \times 7 \times 6 \times 5 \times 4 \times 3 \times 2 \times 1}{5 \times 4 \times 3 \times 2 \times 1} = 8 \times 7 \times 6 = 336$$

If you are comfortable using factorials, you need not write out all the factors. Notice that all the factors in the denominator will be canceled by those in the numerator. This allows you to skip the middle step and just write the remaining factors in the numerator. In this example, because $8 - 5 = 3$ you need write only the first three factors, $8 \times 7 \times 6$.

Example: How many ways can a local political party choose five of its 27 members as delegates to the state convention? Using the permutations formula:

$$P(27, 5) = \frac{27!}{(27-5)!} = \frac{27!}{22!} = 27 \times 26 \times 25 \times 24 \times 23 = 9{,}687{,}600$$

Example: Use the permutation formula to show that there are $n!$ ways of selecting n objects from a set of n objects:

$$P(n, n) = \frac{n!}{(n-n)!} = \frac{n!}{0!} = \frac{n!}{1} = n!$$

Thus far we have assumed that the n objects (from which r objects are selected to form a permutation) are all distinct. If they are not, we cannot determine how many *different* permutations there are without making a change in the formula. Consider, for example, the letters in the word "topper." If we put subscripts on the two p's by labeling them p_1 and p_2, then there are 6! = 720 ways of arranging the six letters (i.e., selecting all six letters). With the subscripts we can distinguish between "to$p_1p_2$er" and "to$p_2p_1$er." Without them we cannot. Because each pair of words that differ only by the order of the two p's are indistinguishable without the subscripts, there are really only 720/2 = 360 different ways of ordering the letters in "topper." If a word had three letters that were the same, then the number of different permutations would be reduced by a factor of six because three letters can be arranged six different ways (i.e., 3! = 6). If the same word had another letter that appeared twice, then the number of different permutations would be reduced by an additional factor of two. More generally, the total number of different permutations of n objects of which r_1 are the same, r_2 others are the same, . . . , and r_j others are the same is

$$\frac{n!}{r_1! \times r_2! \times \ldots \times r_j!}$$

Of course, if all the objects are different, then each of the $r_i! = 1! = 1$ and the equation reduces to $n!$.

Example: How many different ways can the letters in the word "statistics" be arranged? Using the above formula:

$$\frac{10!}{3! \times 3! \times 2! \times 1! \times 1!} = \frac{10 \times 9 \times 8 \times 7 \times 6 \times 5 \times 4 \times 3 \times 2 \times 1}{3 \times 2 \times 1 \times 3 \times 2 \times 1 \times 2 \times 1 \times 1 \times 1} = 50{,}400$$

I need not have written all the "1"s because they do not affect the calculation, but I wanted to show you that I accounted for all 10 letters in the word.

This brings us to *combinations*. Suppose someone is taking a political survey and must select three families to interview from the 10 that live on a particular block. As a permutation, we know there are $10 \times 9 \times 8 = 720$ different ways of selecting the three families. The interviewer, however, probably does not care about the order in which the families are selected. To the interviewer, (F_1, F_2, F_3) is the same as (F_1, F_3, F_2), which is the same as (F_3, F_2, F_1), which is the same as any of the other three orderings of the three families. Thus, rather than 720 different ways of choosing the three families from the 10 on the block, there are only $720/3! = 720/6 = 120$. Of course, the 3! represents the number of ways the three selected families can be ordered. More generally, we must divide the number of permutations for r objects selected from a set of n objects by $r!$ to obtain the number of combinations. We denote the number of combinations of r objects taken from a set of n objects as

$$\binom{n}{r} = \frac{n!}{r! \times (n-r)!}$$

Note: In this context, we can think of a "combination" as a "subset." When we want to know how many combinations of r objects there are in a set of n objects we are really asking how many subsets of r objects can be formed from a set of n objects if we do not care about the order in which any of the r objects became elements of the subset.

Example: How many ways can a voter select three candidates from a field of seven? Using the formula for combinations:

$$\binom{7}{3} = \frac{7!}{3! \times (7-3)!} = \frac{7!}{3! \times 4!} = \frac{7 \times 6 \times 5}{3 \times 2} = 35$$

Notice that I immediately canceled the 4!, the larger of the two factors in the denominator, to save a bit of writing.

Example: The local political party from the earlier example probably does not care about the order in which it chooses five of its 27 members as delegates. If so, then the number of combinations of five delegates it can send to the state convention is

$$\binom{27}{5} = \frac{27!}{5! \times (27-5)!} = \frac{27!}{5! \times 22!} = \frac{27 \times 26 \times 25 \times 24 \times 23}{5!} = \frac{9{,}687{,}600}{120} = 80{,}730$$

2. LIMITS AND CONTINUITY

In this chapter we examine two concepts that are fundamental to an understanding of calculus: limits and continuity.

2.1 The Concept of a Limit

Calculus primarily involves two complementary operations: differentiation and integration. *Differentiation* is a way to calculate an instantaneous rate of change, and it allows us to find maximum values (maxima) or minimum values (minima) of a function if they exist. To illustrate these concepts, consider Olympic champion Carl Lewis running the 100 meters. If he finishes the race in 10.0 seconds we can calculate his speed (rate of change) to be 36 kilometers per hour. This speed, however, is his *average* speed over the entire 100 meters. We know he had a minimum speed of 0 km/hr at the start of the race and a maximum somewhere in the second half of the race, which he may or may not have maintained through the finish line. What if we want to know his speed at the 50-meter mark? A first guess might be his average speed of 36 km/hr, but we know his speed varied considerably from this value during the race. For our second guess we might record his time at the 25 and 75-meter marks and find his average time over that 50 meters. This guess will be closer than the first but still not completely accurate. We could try again by taking times at the 45 and 55-meter marks, then averaging. Again, this third guess will be better than the second, but still a bit off. To get more and more accurate estimates of Lewis's speed at the 50-meter mark, we continue to take measurements of smaller and smaller intervals.

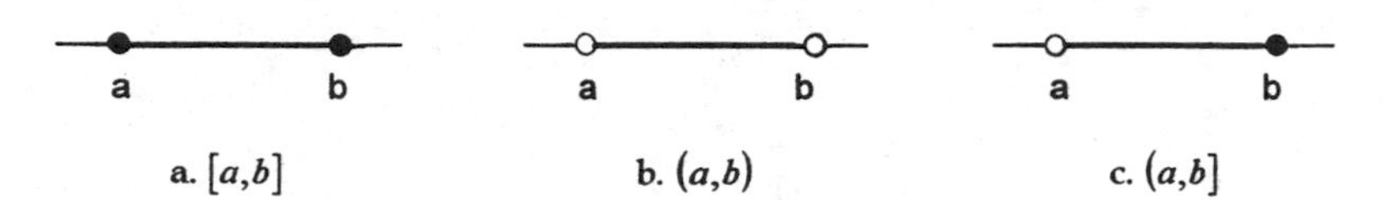

a. $[a,b]$ b. (a,b) c. $(a,b]$

Figure 2.1. Representations of Open and Closed Intervals

At some point this method fails because of the limitations of our measuring instruments; for example, we may not be able to accurately measure time intervals of less than one thousandth of a second. Nevertheless, the example illustrates the underlying notion of how a *limit* is used to calculate instantaneous speed.

Before proceeding to a more rigorous discussion of limits, here is some notation for intervals. For two numbers $a < b$,

1. $[a, b] = \{x \mid a \leq x \leq b\}$ is a closed interval
2. $(a, b) = \{x \mid a < x < b\}$ is an open interval
3. $(a, b] = \{x \mid a < x \leq b\}$ is a half-open, half-closed interval

A bracket next to either a or b indicates that x can take on that value and the interval is *closed* at that endpoint. A parenthesis next to either endpoint indicates that x cannot take on that value and the interval is *open* at that endpoint. Number 3 shows that it is possible for an interval to be open at one endpoint and closed at the other. These intervals are represented graphically in Figure 2.1. The open endpoint is represented by an open circle and the closed endpoint by a filled circle.

The example with Carl Lewis demonstrates the idea behind a limit. Specifically, if you want to know the value of a function at some point, but cannot simply measure it there, measure progressively smaller intervals around that point until you "close in" on the value of the function at the point of interest.

Example: Consider the function $f(x) = x^2$ on the interval $[1, 3]$. We would like to know the limit of the function as x approaches the value 2, or, using the notation for limits, $\lim_{x \to 2} f(x)$. As in the Carl Lewis example, we will take "measurements" of the function at the endpoints of the interval and work our way in to $x = 2$.

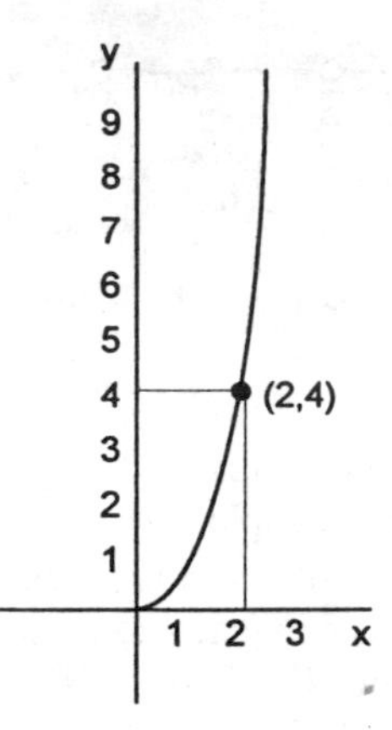

Figure 2.2. Graph of $\lim_{x \to 2} x^2 = 4$

x	$f(x)$
1.0	1
1.5	2.25
1.75	3.06
1.9	3.61
1.99	3.96
1.999	3.996
1.9999	3.9996

x	$f(x)$
3.0	9
2.5	6.25
2.25	5.06
2.1	4.41
2.01	4.04
2.001	4.004
2.0001	4.0004

As you can see from either the set of data points or the graph of the function in Figure 2.2, we can get as close to 4 as we want by choosing x closer and closer to 2. Thus, $\lim_{x \to 2} f(x) = 4$. It may appear that we could have saved some effort by simply entering 2 into the function and calculating the result, that is, $2^2 = 4$. Although this would have given us the correct limit for this function, we really do not care what the value of the function is *at* $x = 2$; rather, we want to know what value the function *approaches* as $x \to 2$. In fact, the function may not even be defined at the point of interest.

Example: What is $\lim_{x \to 2} x^2(x-2)/(x-2)$? Here we cannot just enter 2 for x in the function because it would make the denominator equal to 0. Thus the function is undefined at $x = 2$. The limit, however, does exist. For purposes of the limit, we can cancel $(x-2)$ from the numerator and denominator and find $\lim_{x \to 2} x^2$, which we know to be equal to 4. We must remember, however, that the function is undefined at $x = 2$ despite the existence of the limit. (*Note:* the graph of this function is the same as for $f(x) = x^2$ except that it has a "hole" in it at $x = 2$.)

A more formal definition of a limit can take the following form: The function f is said to have a limit, L, at a point a, written, $\lim_{x \to a} f(x) = L$, if for any neighborhood, $\varepsilon > 0$, of L there exists a neighborhood, $\delta > 0$, about x such that

$$x - \delta < x < x + \delta \Rightarrow L - \varepsilon < L < L + \varepsilon$$

In words, if you choose a value within the specified neighborhood of x, it will yield a value of $f(x)$ that is within the specified neighborhood of L.

Note: $\lim_{x \to a} f(x)$ exists iff both $\lim_{x \to a+} f(x)$ and $\lim_{x \to a-} f(x)$ exist and are equal. The difference between these two *one-sided limits* is simply whether we are approaching a from the right (+) or the left (–) along the x-axis. Put another way, limits are unique. (These one-sided limits are sometimes referred to as right- or left-handed limits.)

Example: Consider $\lim_{x \to 0} f(x)$ for this simple function and its graph in Figure 2.3:

$$f(x) = \begin{cases} 1, & \text{if } x > 0 \\ 0, & \text{if } x = 0 \\ -1, & \text{if } x < 0 \end{cases}$$

If we approach 0 from the right, $\lim_{x \to 0+} f(x) = 1$, but if we approach from the left, $\lim_{x \to 0-} f(x) = -1$. Because the two one-sided limits do not equal each other, this function does not have a limit at $x = 0$. [Remember, for purposes of determining if a limit exists we do not care that $f(x) = 0$.]

2.2 Infinite Limits and Limits at Infinity

Although it is not a specific number, we do consider infinity to be an acceptable value for a limit. Consider the function $f(x) = 2/(x-1)^2$. What

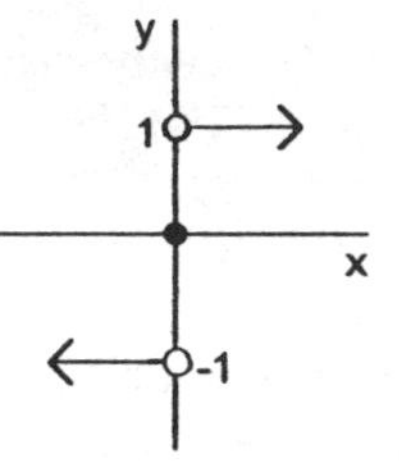

Figure 2.3. Example of a Function With no Limit at $x = 0$

happens to $f(x)$ as x approaches 1? You should plot a few points to see that the graph of the function appears as in Figure 2.4. As you can see from the graph, the values for $f(x)$ continue to get larger and larger as x approaches 1. This should not be a surprise. As $x \to 1$ the value of $(x-1)^2$ gets smaller and smaller, which yields larger and larger values of $f(x)$. Thus we say that $\lim_{x \to 1} f(x) = +\infty$, or that the limit of $f(x)$ grows without bound.

Note: As mentioned in Chapter 1, it is not always easy to determine the overall shape of a function's graph from just a few points. Although you should have an idea of the shape of the graph before you plot any points, this knowledge comes primarily from practice and you should be able to get information on the characteristics of the graph just from looking at the function. For example, in the above example you should know that the graph will have no negative values for $f(x)$ because the numerator is a positive constant and the denominator must be positive because $(x-1)$ is squared. You should see that the larger x gets, positive or negative, $f(x) \to 0$. As previously discussed, you should also see that as $x \to 1$ the values of $f(x)$ grow without bound. Plotting points shows how quickly the function increases or decreases in value, and, for this function, where it crosses the y-axis.

Some functions have limits even though we allow x to grow without bound. Consider again the graph in Figure 2.4 and its associated function, $f(x) = 2/(x-1)^2$. It should be clear from the graph that as $x \to +\infty$, $f(x)$ gets closer and closer to 0; that is, $\lim_{x \to +\infty} 2/(x-1)^2 = 0$. More generally, and more formally, $\lim_{x \to \infty} f(x) = L$ if for each positive number ε there exists a number M such that $L - \varepsilon < f(x) < L + \varepsilon$ when $x > M$. Regarding the graph in Figure 2.4, this means we can get $f(x)$ within some small distance ε of 0 by choosing a large enough value for x.

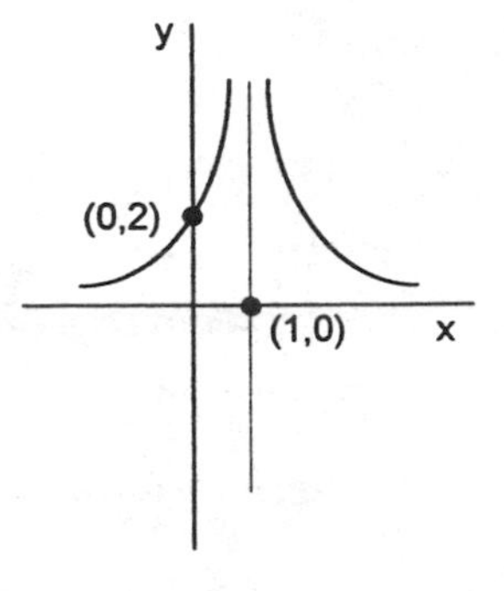

Figure 2.4. $\lim_{x \to 1} 2/(x-1)^2 = +\infty$

Notice that as x approaches negative infinity, $f(x)$ also approaches 0; that is, $\lim_{x \to -\infty} 2/(x-1)^2 = 0$. Although the limit has the same value as x approaches either positive or negative infinity, there is a major difference from previous examples. Specifically, because we are allowing x to go to either positive or negative infinity, we are examining the limit from only one direction: from the left for positive infinity and from the right for negative infinity. We can think about this difference in two ways. First, we might simply consider it to be an exception to the requirement that a limit must exist coming from both directions. We can justify such an exception on the grounds that it is impossible to approach infinity from the other direction. We might also view a limit at infinity as a form of one-sided limit. Because the limit can be checked from only one side, we can treat it as a regular limit. It is unlikely that the matter will come up in most situations.

2.3 Properties of Finite Limits

Suppose that $\lim_{x \to a} f(x)$ and $\lim_{x \to a} g(x)$ both exist. Then,

1. $\lim_{x \to a} [f(x) \pm g(x)] = \lim_{x \to a} f(x) \pm \lim_{x \to a} g(x)$; addition and subtraction

2. $\lim_{x \to a} [f(x) \times c] = c \times \lim_{x \to a} f(x)$; multiplication by a constant

3. $\lim_{x \to a} [f(x) \times g(x)] = \left(\lim_{x \to a} f(x)\right)\left(\lim_{x \to a} g(x)\right)$; multiplication

4. $\lim_{x \to a} [f(x)/g(x)] = \lim_{x \to a} f(x) \Big/ \lim_{x \to a} g(x)$, if $\lim_{x \to a} g(x) \neq 0$; division

5. $\lim_{x \to a} [f(x)]^n = \left[\lim_{x \to a} f(x)\right]^n$, if $\lim_{x \to a} f(x) > 0$; exponent rule

Example: If $f(x) = mx + b$, what is $\lim_{x \to a} f(x)$? Using the limit properties we can simplify the function and find the limits of its component parts. Thus

$$\lim_{x \to a} f(x) = \lim_{x \to a} (mx + b) = \lim_{x \to a} mx + \lim_{x \to a} b = m \lim_{x \to a} x + \lim_{x \to a} b$$

Because b is a constant and does not vary as the value of x changes, $\lim_{x \to a} b = b$. It should be readily apparent that $\lim_{x \to a} x = a$. Substituting these values we find that $\lim_{x \to a} mx + b = ma + b$.

Of course, $mx + b$ is the traditional formula for a line, where m is the slope and b is the y-intercept. This example shows that the limit of a line as $x \to a$ is simply the y value at $x = a$. On a practical level, there will be many times when the limit of a simple function will be the value of the function evaluated at the point of interest. With practice, you should be able to determine whether this is the case without extensive initial calculations. Do not forget, however, that the function need not be defined at the point of interest for the limit to exist.

Example: Consider the following limit:

$$\lim_{x \to 3} \sqrt{\frac{2x^3 + 3x + 4}{3x^2 + 2}} = \lim_{x \to 3} \left(\frac{2x^3 + 3x + 4}{3x^2 + 2}\right)^{1/2} = \left(\frac{\lim_{x \to 3} (2x^3 + 3x + 4)}{\lim_{x \to 3} (3x^2 + 2)}\right)^{1/2}$$

$$= \left(\frac{\lim_{x \to 3} 2x^3 + \lim_{x \to 3} 3x + \lim_{x \to 3} 4}{\lim_{x \to 3} 3x^2 + \lim_{x \to 3} 2}\right)^{1/2}$$

$$= \left(\frac{2 \lim_{x \to 3} x^3 + 3 \lim_{x \to 3} x + \lim_{x \to 3} 4}{3 \lim_{x \to 3} x^2 + \lim_{x \to 3} 2}\right)^{1/2}$$

$$= \left(\frac{2\times 3^3 + 3\times 3 + 4}{3\times 3^2 + 2}\right)^{1/2} = \left(\frac{54+9+4}{27+2}\right)^{1/2} = \left(\frac{67}{29}\right)^{1/2}$$

$$\approx 1.52$$

The first step was to replace the radical with an equivalent exponent, because exponents are usually easier to manipulate. Next I used the properties of limits to reduce the complex function to a collection of limits of simpler functions. It was clear that there would be no problem in evaluating these simpler functions at $x = 3$, and I did so to obtain the final numbers.

Note: Always reduce your answers as much as possible. The above example is finished as a decimal approximation. In some circumstances it may be more appropriate to leave the answer as $(67/29)^{1/2}$. Many textbook examples are designed to work out to some "neat" answer. If you switch to a decimal approximation too quickly you may get something that is "messier" and, more important, incorrect.

Example: Consider $\lim_{x\to+\infty} (5x^2+2)/(3x^2+x+4)$. Here we cannot simply substitute for x in the function and calculate the result. It simply does not make sense to talk about "five times infinity squared plus two" and so on. We need to change the form of the function to something that will make more sense to evaluate. To do so we multiply the function by 1 as follows:

$$\lim_{x\to+\infty} \frac{5x^2+2}{3x^2+x+4} = \lim_{x\to+\infty}\left(\frac{5x^2+2}{3x^2+x+4}\times 1\right) = \lim_{x\to+\infty}\left(\frac{5x^2+2}{3x^2+x+4}\times\frac{1/x^2}{1/x^2}\right)$$

$$= \lim_{x\to+\infty} \frac{(5x^2+2)(1/x^2)}{(3x^2+x+4)(1/x^2)} = \lim_{x\to+\infty}\frac{5+2/x^2}{3+1/x+4/x^2} = \frac{\lim\limits_{x\to+\infty} 5 + \lim\limits_{x\to+\infty} 2/x^2}{\lim\limits_{x\to+\infty} 3 + \lim\limits_{x\to+\infty} 1/x + \lim\limits_{x\to+\infty} 4/x^2}$$

The "trick" here was to multiply the numerator and denominator of the function by one over x raised to the highest power in the *denominator*. (It was just a coincidence that this was also the highest power in the numerator.) Doing so will yield a constant for the term in the denominator with the highest power of x and 0 for the other terms [because $\lim_{x\to+\infty}(1/x^n) = 0$ for $n > 0$]. Continuing with the example we find:

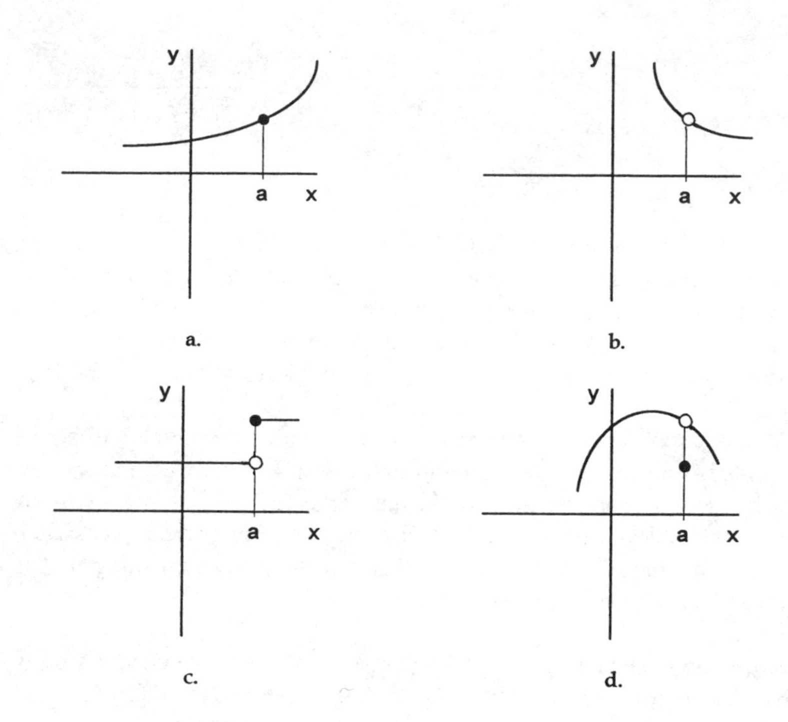

Figure 2.5. Examples of Continuous and Discontinuous Functions

$$\frac{\lim\limits_{x\to+\infty}5+\lim\limits_{x\to+\infty}2/x^2}{\lim\limits_{x\to+\infty}3+\lim\limits_{x\to+\infty}1/x+\lim\limits_{x\to+\infty}4/x^2}=\frac{\lim\limits_{x\to+\infty}5+2\lim\limits_{x\to+\infty}1/x^2}{\lim\limits_{x\to+\infty}3+\lim\limits_{x\to+\infty}1/x+4\lim\limits_{x\to+\infty}1/x^2}=\frac{5+0}{3+0+0}=\frac{5}{3}$$

2.4 Continuity

If we have a function $f(x)$ such that $\lim\limits_{x\to a}f(x)=f(a)$ for every a where $f(x)$ is defined, then we say that $f(x)$ is *continuous*. We can also speak of $f(x)$ being continuous on an interval (e.g., $a<x<b$) and not care whether it is continuous elsewhere. For practical purposes, continuity simply means there are no gaps or "holes" in the function; that is, you can draw the function without lifting your pen from the paper. Consider the four graphs in Figure 2.5. Of the four, only Figure 2.5a is continuous at $x = a$. For a function to be continuous at a point, a, three conditions must be met:

(a) $f(a)$ is defined, (b) $\lim_{x \to a} f(x)$ exists, and (c) $f(a) = \lim_{x \to a} f(x)$. (Which of these conditions are satisfied by each of the graphs in Figure 2.5?)

If $f(x)$ is not continuous at a, it is said to be *discontinuous* at a, or to have a *discontinuity* there. For $f(x)$ to be continuous on an interval, it must be continuous at every point in that interval.

Some Properties of Continuous Functions:

1. If f and g are continuous at point a, then $f + g, f - g$, and $f \times g$ are all continuous at a; f/g is continuous at a iff $g(a) \neq 0$.
2. If f is a polynomial function, then f is continuous for all real values.
3. If $f(x)$ is continuous at $g(a)$ and g is continuous at a, then $f \circ g$: x, the composition of f and g, is continuous at a.

Note: If a function $f(x)$ is continuous at point a, then $\lim_{x \to a} f(x)$ exists (e.g., Figure 2.5a). Conversely, a discontinuity at point a does not imply that $\lim_{x \to a} f(x)$ does not exist. In other words, $\lim_{x \to a} f(x)$ may exist even if $f(x)$ is discontinuous at point a (e.g., Figures 2.5b and 2.5d). Of course, $f(x)$ may be discontinuous at point a and $\lim_{x \to a} f(x)$ may not exist (e.g., Figure 2.5c).

3. DIFFERENTIAL CALCULUS

In Chapter 2 we used the notion of a limit to determine Carl Lewis's speed at the 50-meter mark in a 100-meter race. We knew his speed at the 50-meter mark might not be the same as his average speed because his speed varied during the race. Suppose we consider the distance Lewis has traveled to be a function of time, $f(t)$. Then the instantaneous rate of change of that function is its *derivative* [i.e., the derivative of $f(t)$ is Lewis's speed]. Finding the derivative of a function is called *differentiation*. Differentiation is a very powerful tool that is used extensively in model estimation. Practical examples of differentiation are usually in the form of optimization problems or rate of change problems. You will see examples of both in this chapter.

Let us begin by considering a simple cost curve, $C(x) = 3x + 2$, where x is the level of production of some product. You can consider $3x$ to be costs associated with the production of each unit (variable costs) and the 2 to be costs that are incurred regardless of the number of units produced (fixed costs). Although this cost curve is not very realistic, it provides for an easy introduction to differentiation. (*Note:* I did not use the traditional f to represent the function, nor did I use g or h, the usual alternatives. Some disciplines use special letters to represent particular functions. It should be easy to adapt to any particular usage provided the authors are clear about how letters and symbols are being used and are consistent in using them.) What is the average increase in cost in going from one level of x produced to another? The average increase in cost will be the change in cost from one level to another divided by the change in the amount of the product from one level to another. We can represent the average increase in cost as follows:

$$\frac{\Delta C}{\Delta x} = \frac{\text{change in cost}}{\text{change in production}} = \frac{y_1 - y_0}{x_1 - x_0} = \frac{C(x_1) - C(x_0)}{x_1 - x_0}$$

These are four different ways of representing the same quantity. In the first quotient, delta (Δ) means "the change in." The third quotient uses y as the familiar $y = C(x)$. The subscripts in the last two quotients are used to distinguish different values of a particular variable. Previously, a, b, and c were often used to represent different points along, for example, the x-axis. When several values are involved remembering the variable to which they refer can become difficult. The use of subscripts solves this problem. By convention, the 0 subscript usually refers to the starting point of a series of values. In this example x_0, called "x naught," represents the initial production level.

Returning to the example, we can substitute the formula for $C(x)$ into the final quotient to obtain

$$\frac{(3x_1 + 2) - (3x_0 + 2)}{x_1 - x_0} = \frac{3x_1 - 3x_0}{x_1 - x_0} = \frac{3(x_1 - x_0)}{x_1 - x_0} = 3.$$

You should notice two things from this result. First, because the function is linear, the rate of change is equal to the slope of the line. Second, and also because the cost function is linear, the rate of change is constant. In other words, the costs of production will continue to rise (or fall) at a constant rate regardless of the production level of x.

Now consider a nonlinear cost curve, $C(x) = x^2$. Making the substitutions as before, the average rate of change of $C(x)$ is

$$\frac{\Delta C}{\Delta x}=\frac{x_1^2-x_0^2}{x_1-x_0}=\frac{(x_1+x_0)(x_1-x_0)}{x_1-x_0}=x_1+x_0$$

In this example, the rate of change of the cost depends on the levels of production being considered; that is, it is not constant. Because the rate of change is not constant, we would like to be able to determine the *marginal cost* at each specific level of production. The marginal cost is the change in the cost as we move from one particular level of production to the next. Similarly, marginal revenue (benefits) is the change in revenue from one level of production to the next. To maximize profit, an optimization problem, we find the level of production where marginal costs equal marginal revenues (i.e., where the slope of the cost function equals the slope of the revenue function).

As another rate of change problem, consider Carl Lewis again. Suppose that $d=f(t)$ tells us how much distance Lewis has traveled at a particular point in time. We can use this function to calculate his instantaneous speed at time t:

$$v=\frac{\Delta d}{\Delta t}=\frac{f(t_1)-f(t_0)}{t_1-t_0}$$

where v = velocity, d = distance, and t = time. Although the letters representing the variables have changed from the cost problem, the form of the equation has not.

3.1 Tangents to Curves

A *tangent* (or tangent line) is a line that touches a curve at only one point and is parallel to the curve at that point. Suppose we want to know the slope of the tangent line for a point on the curve formed by the graph of a function. Consider the graph in Figure 3.1. Begin by calculating the slope of the line connecting (x_0, y_0) and (x_1, y_1) using the formula

$$\frac{\Delta y}{\Delta x}=\frac{y_1-y_0}{x_1-x_0}$$

In a sense, the slope of the line connecting these two points is like the average speed we could calculate for Carl Lewis between the start and finish of a race. Just as we obtained more accurate estimates of his speed

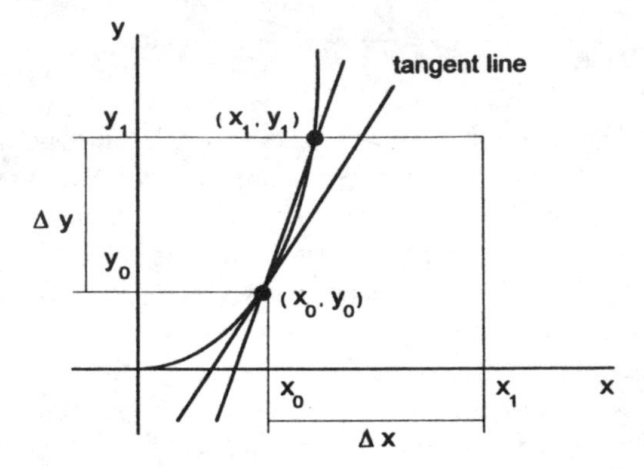

Figure 3.1. Representation of a Tangent Line

at a particular point by examining smaller and smaller intervals around that point, as $x_1 \rightarrow x_0$ the line between (x_0, y_0) and (x_1, y_1) approaches the tangent line at x_0 and the slope of the line approaches the slope of the tangent line. Thus the slope of the tangent line at x_0 is equal to

$$\lim_{x_1 \to x_0} \frac{y_1 - y_0}{x_1 - x_0}$$

Example: If $f(x) = 2x^2 - x$ and $x_0 = 2$, then the slope of the tangent line to $f(x)$ at $x = 2$ is equal to

$$\lim_{x \to 2} \frac{(2x^2 - x) - (2 \times 2^2 - 2)}{x - 2} = \lim_{x \to 2} \frac{2x^2 - x - 6}{x - 2} = \lim_{x \to 2} \frac{(2x + 3)(x - 2)}{x - 2}$$

$$= \lim_{x \to 2} 2x + 3 = 7$$

Sometimes you will see examples that use $\lim_{h \to 0}$, where $h = x_1 - x_0$. This is just another way of stating the same problem. Using this notation for this example we have

$$\lim_{h \to 0} \frac{f(2 + h) - f(2)}{2 + h - 2} = \lim_{h \to 0} \frac{2(2 + h)^2 - (2 + h) - 6}{h}$$

$$= \lim_{h \to 0} \frac{2(4+4h+h^2)-(2+h)-6}{h}$$

$$= \lim_{h \to 0} \frac{2h^2+8h+8-2-h-6}{h}$$

$$= \lim_{h \to 0} \frac{2h^2+7h}{h}$$

$$= \lim_{h \to 0} 2h+7$$

$$= 7$$

Although using $h \to 0$ rather than $x_1 \to x_0$ seems a bit messier, using a small difference from the point of interest (h or Δx) is a standard technique in mathematical proofs.

Consider the cost curve, $C(x) = x^2$, again. Marginal cost, $M(x)$, at x_0 is the limit as $x_1 \to x_0$ (or as $h \to 0$ using the other notation). When

$$\frac{\Delta C}{\Delta x} = x_1 + x_0, \text{ then } M(x_0) = \lim_{\Delta x \to 0} \frac{\Delta C}{\Delta x} = \lim_{x_1 \to x_0} x_1 + x_0 = 2x_0$$

where $M(x_0)$ is the marginal cost at x_0. Notice that I used $\Delta x \to 0$ in the above equation. This is just another way of stating the equation.

We can now formally state the definition of a *derivative* as

$$\lim_{\Delta x \to 0} \frac{f(x+\Delta x)-f(x)}{\Delta x}$$

We could also state the definition using $x_1 \to x_0$ or $h \to 0$.

Example: If $f(x) = x^2$, what is the derivative at $x = 3$? First, determine some important values.

Δx	x_0	$x_1 = x_0 + \Delta x$	y_0	$y_1 = f(x_1)$	$\Delta y = y_1 - y_0$	$\Delta y/\Delta x$
1	3	4	9	16	7	7
.5	3	3.5	9	12.25	3.25	6.5
.01	3	3.01	9	9.0601	.0601	6.01
.001	3	3.001	9	9.006001	.006001	6.001
.0001	3	3.0001	9	9.00060001	.00060001	6.0001

It should be clear that we can get $\Delta y/\Delta x$ as close to 6 as we want by making Δx smaller and smaller. The numbers above show numerically what Figure 3.1 showed graphically. In this example, $x_0 = 3$ and we take five values for x_1 beginning with 4. We calculated the slope, $\Delta y/\Delta x$, of each of the five lines formed by the two points. As you will soon see, we do not need to do such elaborate calculations every time we want to determine a derivative (but it is good to see it at least once).

Many times we want to know the derivative of a function without necessarily calculating the derivative at a particular point. To do so we can define a *derivative function* of f, denoted $f'(x)$, at all points where the derivative exists. Just as we have seen three different ways of stating the definition of a derivative at a point, there are several ways of indicating a derivative function. (We usually just refer to the derivative function as the derivative. If we want to calculate the derivative at a particular point, that point will be indicated and we will obtain a specific value for the function.) Some of the most common symbols for derivatives are

$$f'(x),\ \frac{dy}{dx},\ \frac{d}{dx}[f(x)],\ D_x y,\ \text{and}\ D_x f.$$

In most instances these symbols can be used interchangeably. Why so many ways of saying the same thing? There are two primary reasons. First, you can choose the notational style that is most appropriate for a particular context. This will often make the formula easier to write and to remember. Second, not every discipline uses the same symbols, and one should to be familiar with the most common alternatives.

Note: If a function is differentiable at a point then the function is continuous at that point. Continuity, however, does not imply differentiability. In general, if a curve is "smooth" at a point (i.e., no spikes or corners) it will be differentiable there.

3.2 Differentiation Rules

1. If $f(x) = c$, where c is a constant, then $f'(x) = 0$: The derivative of a constant is zero. A constant function does not change, so Δy, and thus $\Delta y/\Delta x$, equals zero.

2. If $f(x) = mx + b$, then $f'(x) = m$: The derivative of a linear function is the slope of the line, m.

3. $\frac{d}{dx}[c \times f(x)] = c \times \frac{d}{dx}[f(x)]$: The derivative of a constant times a function is equal to the constant times the derivative of the function; that is, you can pull a constant through the differentiation.
4. If $f(x) = x^n$, then $f'(x) = nx^{n-1}$: The derivative of x raised to the power n is n times x raised to the power $n - 1$. This is called the power rule.
5. $(f + g)' : x = f'(x) + g'(x)$: The derivative of the sum of two functions is equal to the sum of the derivatives of the functions.
6. $(f \times g)' : x = [f'(x) \times g(x)] + [f(x) \times g'(x)]$: The derivative of the product of two functions is equal to the derivative of the first function times the second function plus the first function times the derivative of the second function. This is called the product rule.
7. $(f/g)' : x = \frac{[f'(x) \times g(x)] - [f(x) \times g'(x)]}{[g(x)]^2}$, if $g(x) \neq 0$: The derivative of the quotient of two functions is equal to the derivative of the numerator times the denominator minus the numerator times the derivative of the denominator all divided by the square of the denominator, provided the denominator is not equal to zero. This is called the quotient rule.
8. $(f \circ g)' : x = g'(x) \times f'[g(x)]$: The derivative of the composition of two functions is equal to the derivative of the second function times the derivative of the first function evaluated at the second function. This is called the chain rule.
9. $$f(x) = e^x \Rightarrow f'(x) = e^x$$ $$f(x) = a^x \Rightarrow f'(x) = (\ln a) \times a^x$$
10. $$f(x) = \ln x \Rightarrow f'(x) = \frac{1}{x}$$ $$f(x) = \log_a x \Rightarrow f'(x) = \frac{1}{x \times \ln a}$$

The first eight are standard rules that apply to most functions. (The numbers identifying the rules have no inherent significance, but I will use them as a shorthand way of referring to them.) Social science estimation procedures often use logarithmic and exponential functions so I also include rules for common forms of those functions.

Examples: Find the derivatives for the following examples

1. $f(x) = x^4$: Using the power rule (4): $f'(x) = 4x^{4-1} = 4x^3$.
2. $f(x) = 5x^3$: Here, think of $f(x) = 5 \times g(x)$, where $g(x) = x^3$, then use rules 3 and 4: $\frac{d}{dx}(5x^3) = 5 \times \frac{d}{dx}(x^3) = 5 \times 3x^2 = 15x^2 = f'(x)$.
3. $f(x) = 3x^4 + 5x^3 + 2x^2 + 7x + 4$: For this example, first use rule 5 to recognize that we can treat each term of the function as a separate function, take their derivatives using rules 3 and 4, then add the derivatives of each term to obtain the derivative of the sum:

$$\begin{aligned} f'(x) &= \frac{d}{dx}(3x^4) + \frac{d}{dx}(5x^3) + \frac{d}{dx}(2x^2) + \frac{d}{dx}(7x) + \frac{d}{dx}4 \\ &= 3 \times 4x^3 + 5 \times 3x^2 + 2 \times 2x^1 + 7x^0 + 0 \\ &= 12x^3 + 15x^2 + 4x + 7 \end{aligned}$$

Notice that both rules 3 and 4 were applied in the second step. Notice also that I left the exponents on $2x^1$ and $7x^0$ in the second step. This was so that you would see I was applying the power rule properly. Finally, rule 1 was also used in this example: The derivative of a constant, here 4, is zero.

4. $f(x) = (2x^2 + x)(3x + 1)$: Here, begin with the product rule (6), then use the rules from previous examples. Think of $g(x) = 2x^2 + x$ and $h(x) = 3x + 1$. Then

$$f'(x) = (g \times h)' : x$$

$$= g'(x) \times h(x) + g(x) \times h'(x)$$

$$= \left(\frac{d}{dx}(2x^2 + x)\right) \times (3x + 1) + (2x^2 + x) \times \left(\frac{d}{dx}(3x + 1)\right)$$

$$\begin{aligned} &= (4x + 1) \times (3x + 1) + (2x^2 + x) \times 3 \\ &= (12x^2 + 4x + 3x + 1) + (6x^2 + 3x) \\ &= 18x^2 + 10x + 1 \end{aligned}$$

When you first start using these rules you should write out each step. As you become more comfortable using the rules you can begin to combine steps as in this example. You may have wondered if one could simply multiply the two factors and then take the derivative of the result. The example was intended to illustrate the use of the product rule, but we can now multiply the factors to check the previous answer:

$$\begin{aligned} f(x) &= (2x^2 + x) \times (3x + 1) \\ &= 6x^3 + 2x^2 + 3x^2 + x \\ &= 6x^3 + 5x^2 + x \\ f'(x) &= 18x^2 + 10x + 1 \end{aligned}$$

In practice, unless you are asked to use a particular rule, use whichever rules make the calculations easier.

5. $f(x) = (x + 1)/(x^2 - 1)$: Here we will use the quotient rule. Think of $g(x) = x + 1$ and $h(x) = x^2 - 1$. Then

$$f(x) = (g/h): x$$

$$f'(x) = \frac{g'(x) \times h(x) - g(x) \times h'(x)}{[h(x)]^2} = \frac{(1) \times (x^2 - 1) - (x + 1) \times (2x)}{(x^2 - 1)^2}$$

$$= \frac{(x^2 - 1) - (2x^2 + 2x)}{x^4 - 2x^2 + 1} = \frac{x^2 - 1 - 2x^2 - 2x}{x^4 - 2x^2 + 1}$$

$$= \frac{-x^2 - 2x - 1}{x^4 - 2x^2 + 1} = \frac{-(x^2 + 2x + 1)}{x^4 - 2x^2 + 1}$$

$$= \frac{-(x + 1)(x + 1)}{(x + 1)(x - 1)(x + 1)(x - 1)} = \frac{-1}{(x - 1)^2}$$

If you had stopped at $\dfrac{-x^2 - 2x - 1}{x^4 - 2x^2 + 1}$ you would not have seen the additional cancellations that were possible. While you should simplify your answer as much as possible, it may not always be obvious how to do so. As in the previous example, I could have factored $x + 1$ from the numerator and denominator at the start and used the chain rule on $(x - 1)^{-1}$. Of course, this probably did not occur to you because I have not given you an example of the chain rule. Let me do so now.

6. $f(x) = (x - 1)^{-1}$: Think of $f(x)$ as being composed of two functions. Specifically, let $h(x) = x - 1$ and $g(y) = y^{-1}$. The introduction of the new variable letter y is just to help in distinguishing the two functions. Now apply the chain rule as follows:

$$
\begin{aligned}
&f(x) = g \circ h : x \\
&h(x) = x - 1 \\
&g(y) = y^{-1} \\
&f'(x) = (g \circ h)' : x = h'(x) \times g'[h(x)] \\
&h'(x) = 1 \\
&g'(y) = -y^{-2} \\
&f'(x) = 1 \times [-(x - 1)^{-2}] = \frac{-1}{(x - 1)^2}
\end{aligned}
$$

This is the same answer obtained in the previous example, but by using a different method. The key concept in using the chain rule is in evaluating the derivative of the first function evaluated at the second, $g'[h(x)]$ in this example. In essence, to evaluate $g'[h(x)]$, I substituted $h(x) = x - 1$ for y in $g'(y) = -y^{-2}$. This concept can be difficult to grasp at first so here is another example.

7. $f(x) = (2x^3 + x^2 - 4x)^5$: This example is a relatively simple polynomial raised to the fifth power. Here, think of the two component functions as $h(x) = 2x^3 + x^2 - 4x$ and $g(y) = y^5$. Then

$$
\begin{aligned}
&f'(x) = (g \circ h)' : x \\
&f'(x) = h'(x) \times g'[h(x)] \\
&g'(y) = 5y^4 \\
&h'(x) = 6x^2 + 2x - 4 \\
&f'(x) = (6x^2 + 2x - 4) \times 5(2x^3 + x^2 - 4x)^4 \\
&\qquad = 10(3x^2 + x - 2)(2x^3 + x^2 - 4x)^4
\end{aligned}
$$

Little could be done to reduce the final answer except to factor a 2 from the first term and multiply it with the 5. One *could* do all the multiplications, but it should be rare that such effort is required.

8. $f(x) = e^{1/x^3}$: The final example of this section uses e, the base for natural logarithms. This example will require the use of the chain rule. Following the procedures from the previous two examples:

$$f(x) = g \circ h : x$$
$$g(y) = e^y$$
$$h(x) = \frac{1}{x^3} = x^{-3}$$
$$f'(x) = (g \circ h)' : x = h'(x) \times g'[h(x)]$$
$$g'(y) = e^y$$
$$h'(x) = -3x^{-4}$$
$$f'(x) = (-3x^{-4}) \times e^{x^{-3}} = \frac{-3e^{x^{-3}}}{x^4} = -3x^{-4}e^{x^{-3}}$$

The final form of the answer is a matter of choice. From the last line it is easy to see that the derivative of an exponential function of this form is just the original function multiplied by the derivative of the power to which e is raised.

We can also take the derivative of the derivative of a function. This is called a *higher order derivative*. Returning to the Carl Lewis example, if his distance from the starting line is $f(t)$, then his instantaneous speed at time t is found by taking the derivative, $f'(t)$. To find Lewis's *acceleration* at time t, we take the derivative of $f'(t)$. Like derivatives, there are several ways of representing higher order derivatives. A few of the more common representations for a *second derivative* (the derivative of a derivative) are

$$f''(x) = \left(\frac{d}{dx}\right)\left(\frac{dy}{dx}\right) = \frac{d^2y}{dx^2} = \frac{d^2f}{dx^2} = D_x^2 y = f^2(x) = f_{xx}$$

Again, the particular symbol chosen will depend on context and ease of use. I will primarily use $f''(x)$ to represent second derivatives in this chapter. In the chapter on matrices (Chapter 6), I will be more likely to use

f_{xx}. Notice the 2 in the above symbols. It indicates that the symbols represent a second derivative. For a third derivative a 3 would be used, and so on. For a polynomial function, one can continue to take higher order derivatives until the term with the highest power of x (or the variable of interest) is reduced to zero.

3.3 Extrema of Functions

When we speak of *extrema* of functions we are referring to where the function takes on extreme values, highest or lowest, maximum or minimum. Such extrema can be of two types: *relative* (or local) and *absolute* (or global). If a point is an absolute maximum (or minimum) it means no other value of x will yield a greater (or smaller) value of $f(x)$. If a point is a local maximum (or minimum) it indicates that no other value of x in the neighborhood of the extremum will yield a larger (or smaller) value of $f(x)$, though other values of x outside this neighborhood may.

Consider the function graphed in Figure 3.2. As x gets very large or very small, $f(x)$ goes to $-\infty$. In the region we see in the graph, however, the function exhibits some interesting behavior. At x_2 the function has a smaller y value than at any other point in the immediate neighborhood. Thus there is a local minimum at x_2. Notice also that the function has local maximum at both x_1 and x_3. Because $f(x)$ gets smaller for x values less than x_1 and greater than x_3, the function must have an absolute maximum at one of the two points [or both if $f(x_1) = f(x_3)$].

Before explaining why we want to find the extrema of functions, I will show how to find them. First, if the function has an extremum at x_0 (i.e., $f(x_0)$ is a max or min, local or global), and if $f'(x_0)$ exists, then $f'(x_0) = 0$. This is a very important point. It tells us that, at extreme points, the slope of the line tangent to the function is equal to zero. We can see this visually by looking at Figure 3.2 again. Visualize how the slope of the tangent line decreases from a steep positive slope as we move toward x_1 from the left. It continues to level off until it is horizontal at x_1. A horizontal tangent line means the slope of the tangent line is zero and the rate of change of the function at that point is zero. Initially the slope of the tangent line becomes increasingly negative as we continue along the curve toward x_2. The slope eventually stops becoming more negative and begins to level off once again as it gets close to x_2. At x_2 the tangent line is again horizontal and its slope is equal to zero. After passing x_2 the slope begins to increase rapidly, reaching a peak somewhere between x_2 and x_3, after which it once again

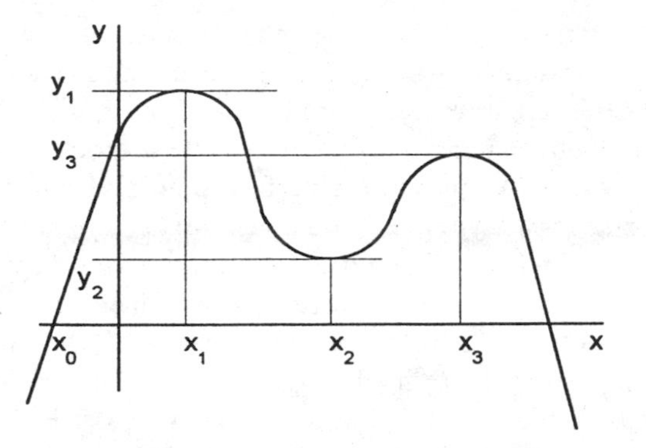

Figure 3.2. Relative and Absolute Extrema

begins to level off. Again, the tangent line is horizontal at x_3 and its slope is zero.

To find the extreme points of a function, we find the derivative of the function, set it equal to zero, solve the resulting equation, and check the points to determine whether they are maxima or minima. (*Note:* It is possible that such points are neither maxima nor minima, but rather points of inflection. Some texts use the term *critical points* rather than "extrema" to allow for this third possibility.)

Example: Find any local extrema of $f(x) = x^3 - x$. Following the above procedure, first find the derivative of $f(x)$, then set it equal to zero and solve for x:

$$f(x) = x^3 - x$$

$$f'(x) = 3x^2 - 1$$

$$3x^2 - 1 = 0 \Rightarrow 3x^2 = 1 \Rightarrow x^2 = \frac{1}{3} \Rightarrow x = \pm\frac{1}{\sqrt{3}}$$

We now enter these two values into $f(x)$ to determine the y-axis values for the points we believe to be extreme points:

$$f\left(\frac{1}{\sqrt{3}}\right) = -\frac{2\sqrt{3}}{9} \quad \text{and} \quad f\left(-\frac{1}{\sqrt{3}}\right) = \frac{2\sqrt{3}}{9}$$

Knowing these values, and given the relative simplicity of $f(x)$, we can make a good guess that $(-1/\sqrt{3}, 2\sqrt{3}/9)$ is a relative maximum and $(1/\sqrt{3}, -2\sqrt{3}/9)$ is a relative minimum.

Note: A polynomial function, $f(x)$, will have a number of *roots* (not necessarily all distinct) equal to the highest power of x in the function. These roots come in two types, real and imaginary. Imaginary roots involve some form of $\sqrt{-1}$ and we will not deal with them here except to note they always come in pairs. The number of real roots a function has tells us how many times the graph of the function crosses the x-axis. Knowing the sign and power of the term in $f(x)$ with the highest power and the number of real roots of the function allows us to make a good guess as to which critical points are maxima and which minima. In the above example, we know that $f(x) \to \infty$ as $x \to \infty$ and that $f(x) \to -\infty$ as $x \to -\infty$ because the largest power of x in $f(x)$ is 3 and the sign of the term is positive. Thus $f(x)$ will have either one or three real roots (because imaginary roots come in pairs). We can find these roots by factoring $f(x)$. Factoring, we find $f(x) = x(x + 1)(x - 1)$. Because we are looking for places the function crosses the x-axis, of necessity, this means we want to know where $f(x) = 0$. It should be clear that $f(x) = 0$ when any term of the factored $f(x)$ is equal to zero, so we set each term equal to zero and solve for x. By doing so for this example we find that $f(x)$ crosses the x-axis at $x = -1$, 0, and 1. Knowing that we have three real roots tells us there must be two "bumps" in the graph of the function: one between -1 and 0, and one between 0 and 1. As we approach -1 from the left the values of $f(x)$ are negative. At $x = -1, f(x) = 0$. Between -1 and 0 the values of $f(x)$ are positive. Therefore, somewhere between -1 and 0 there must be a maximum value for $f(x)$. This is the value we calculated earlier. Specifically, it occurs at $(-1/\sqrt{3}, 2\sqrt{3}/9)$. Using similar logic, we deduce that $(1/\sqrt{3}, -2\sqrt{3}/9)$ must be a minimum.

If $f(x)$ is very complex or is not easily factored, or if we just do not know its shape, we need a more rigorous method of determining whether a critical point is a maximum or minimum. There are two tests for doing so.

The easier of the two is the *second derivative test*. As its name suggests, begin by calculating $f''(x)$, the second derivative of $f(x)$. Then evaluate $f''(x)$ at the critical point being tested; that is, enter the value of the critical point, x_0, into the equation for $f''(x)$. There are three possible outcomes: (a) if $f''(x_0) < 0$, then x_0 is a relative maximum; (b) if $f''(x_0) > 0$, then x_0 is a relative minimum; (c) if $f''(x_0) = 0$, then the test is inconclusive (x_0 could be a maximum, minimum, or point of inflection) and we must try the first derivative test.

In the *first derivative test,* examine the sign of $f'(x)$ in a neighborhood around the critical point, x_0. There are three possible outcomes: (a) if for $x < x_0, f'(x) > 0$ and for $x > x_0, f'(x) < 0$, then x_0 is a relative maximum; (b) if for $x < x_0, f'(x) < 0$ and for $x > x_0, f'(x) > 0$, then x_0 is a relative minimum; (c) if the sign of $f'(x)$ does not change as we move from one side of the critical point to the other, then x_0 is a point of inflection.

Note: There is no test for absolute maxima and minima. If we know the shape of the function, we should know whether absolute extrema exist. For example, we can tell the graph in Figure 3.2 has an absolute maximum just by looking at it. Whichever of the relative maxima is the largest is the absolute maximum; that is, the larger of $f(x_1)$ and $f(x_3)$. If we are examining the function on a finite interval then we are more likely to have absolute extrema. On a finite interval we must also check the endpoints. Endpoints will not usually be critical points but may have more extreme values than any other point in the interval. For example, in Figure 3.2 $f(x_0) < f(x_2)$ making $(0, f(x_0))$ the absolute minimum on the finite interval $[x_0, x_3]$. Some functions do not have extrema except at the endpoints of a finite interval (e.g., a linear function).

Example: Find the maxima and minima, if any, of $f(x) = x^3 - 3x - 4$. Before beginning our calculations, you should see from the function itself, specifically the sign and power of the first term, that there will not be any absolute extrema. This function will cross the x-axis in either one or three places (depending on whether it has one or three real roots), which means there will be at most one relative minimum and one relative maximum. Knowing all this before we start will help to guard against obtaining a wrong answer later. Turning to the calculations,

$$
\begin{aligned}
f(x) &= x^3 - 3x - 4 \\
f'(x) &= 3x^2 - 3 = 3(x^2 - 1) = 3(x + 1)(x - 1) \\
x &= -1, 1
\end{aligned}
$$

The critical points of this function are at (–1, –2) and (1, –6). Next determine whether these critical points are relative extrema and, if so, of what type. From the previous discussion, you can probably guess that (–1, –2) is a relative maximum and (1, –6) is a relative minimum. From the second derivative test we find

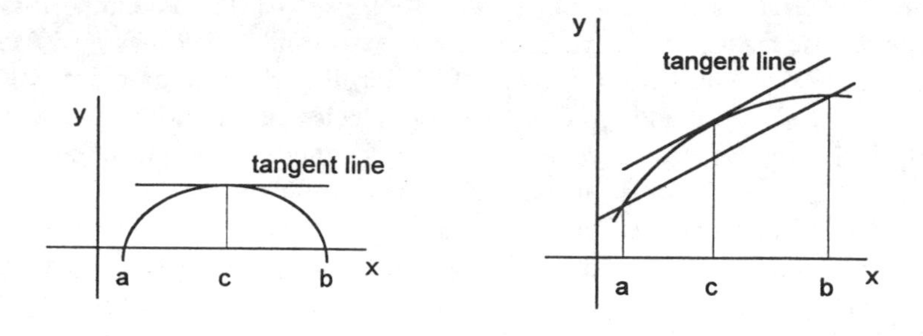

a. Rolle's Theorem

b. Mean Value Theorem

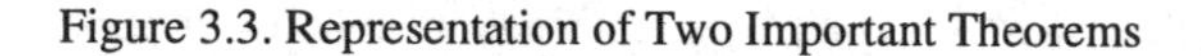
Figure 3.3. Representation of Two Important Theorems

$$f''(x) = 6x$$
$$f''(-1) = -6 < 0$$
$$f''(1) = 6 > 0$$

Because $f''(-1) < 0$, $(-1, -2)$ is a relative maximum, and because $f''(1) > 0$, $(1, -6)$ is a relative minimum, as expected. For practice, also apply the first derivative test to these critical points. Begin with $(-1, -2)$ and check the sign of $f'(x)$ on either side of $x = -1$. Remembering that $f'(x) = 3x^2 - 3$, we find $f'(-1.5) = 3.75 > 0$ and $f'(-.5) = -2.25 < 0$, which tells us that $(-1, -2)$ is a relative maximum. Using the same procedure with $(1, -6)$ we find that $f'(.5) = -2.25 < 0$ and $f'(1.5) = 3.75 > 0$, which tells us that $(1, -6)$ is a relative minimum, confirming the results of the second derivative test and our expectations.

Rolle's Theorem is a more formal statement of some of the previous discussion. Specifically, Rolle's Theorem tells us that, for a function $f(x)$ that is defined and continuous on the closed interval $[a, b]$ and that is differentiable on the open interval (a, b), if $f(a) = f(b) = 0$, then there must be at least one number c between a and b where $f'(c) = 0$. Less formally, if $f(x)$ crosses the x-axis at two points, there must be a relative extremum between those two points, assuming $f(x)$ is continuous and differentiable on the interval between the two points. This is illustrated in Figure 3.3a.

The *Mean Value Theorem* is a generalization of Rolle's Theorem. It says that, for $f(x)$ continuous on $[a, b]$ and differentiable on (a, b), there is some number c for which $f'(c) = [f(b) - f(a)]/(b - a)$; that is, the slope of the tangent line at $x = c$ is equal to the slope of the line between a and b. This is illustrated in Figure 3.3b.

Rolle's Theorem and the Mean Value Theorem are useful in finding the real roots and extrema of a function.

4. MULTIVARIATE FUNCTIONS AND PARTIAL DERIVATIVES

Thus far we have considered functions with only one independent variable. Such functions can be represented graphically in two dimensions: one for the independent variable (plotted on the x-axis) and the second for the value of the function (the dependent variable, plotted on the y-axis). We can also construct functions consisting of two or more independent variables. A function with two independent variables can be represented by a three dimensional graph. The two independent variables are plotted on the xy-plane while the value of the function is plotted on the z-axis. Functions with more than one independent variable are called *multivariate functions*. The notation for multivariate functions is a simple extension of the notation for bivariate functions: A function in two variables, x and y, can be represented as $f(x, y)$. In addition, just as we sometimes refer to $y = f(x)$, we can also say $z = f(x, y)$. A function in three variables might be represented as $w = f(x, y, z)$, and so on.

4.1 Partial Derivatives

Suppose $f(x, y)$ is a function in two variables. We want to determine the rate of change for several points. Even for very simple functions, however, it is not possible to calculate a single number that represents the rate of change at a particular point. This is because the rate of change of the function depends on the direction from which one approaches the base point. (To illustrate the point, think of climbing to the summit of a steep mountain ridge. If you approach the summit along the ridge line the climb will not be too steep. If you approach the summit perpendicular to the ridge line the climb will be very steep.) We can, however, get a good approximation of the total derivative of the function by calculating its partial derivatives.

Partial derivatives, often just called "partials," involve taking the derivatives of parts of the function. A partial derivative can be calculated for each independent variable in the function. For $f(x, y)$, the partial derivative with respect to x is

$$\frac{\partial f(x, y)}{\partial x} = \lim_{h \to 0} \frac{f(x+h, y) - f(x, y)}{h}$$

(Note the similarity to the earlier definition for the derivative of a function. Note also the new symbol, ∂, used to represent partial derivatives.) Similarly, the partial derivative of $f(x, y)$ with respect to y is

$$\frac{\partial f(x, y)}{\partial y} = \lim_{h \to 0} \frac{f(x, y+h) - f(x, y)}{h}$$

Calculating partial derivatives is no more difficult than calculating derivatives for functions of one independent variable. This is because in calculating the partial derivative of a function for a particular variable, all other variables in the function are treated as constants. Understanding this last point is often the most difficult part of calculating partial derivatives.

Example: Calculate $\partial f(x, y)/\partial x$ and $\partial f(x, y)/\partial y$ for $f(x, y) = 2x^2 + y^2$:

$$\frac{\partial f(x, y)}{\partial x} = \frac{\partial(2x^2 + y^2)}{\partial x} = 2 \times 2x + 0 = 4x$$

In calculating $\partial f(x, y)/\partial x$, we treat all appearances of the variable y as constants. Thus the term y^2 is considered a constant and, because the derivative of a constant equals zero, it disappears when we take the partial derivative with respect to x. Now calculate the partial derivative with respect to $y : \partial f(x, y)/\partial y = 0 + 2y = 2y$. Here, all appearances of the variable x are treated as constants. In calculating $\partial f(x, y)/\partial y$ the term $2x^2$ is treated as a constant and becomes zero when we take the partial derivative with respect to y.

Just as we can take second order derivatives, we can also take *second order partial derivatives*. The second order partial derivative of $f(x, y)$ with respect to x and the second order partial derivative with respect to y are represented as follows:

$$\frac{\partial^2 f(x, y)}{\partial x^2} = \frac{\partial}{\partial x}\left(\frac{\partial f(x, y)}{\partial x}\right); \quad \frac{\partial^2 f(x, y)}{\partial y^2} = \frac{\partial}{\partial y}\left(\frac{\partial f(x, y)}{\partial y}\right)$$

These representations can become a bit cumbersome, so these partials are sometimes represented as f_{xx} and f_{yy} (just as the first order partials might be represented as f_x and f_y). We can also calculate *mixed partials* (sometimes called cross partials), which can be represented as

$$\frac{\partial^2 f(x,y)}{\partial x \partial y} = \frac{\partial}{\partial x}\left(\frac{\partial f(x,y)}{\partial y}\right); \frac{\partial^2 f(x,y)}{\partial y \partial x} = \frac{\partial}{\partial y}\left(\frac{\partial f(x,y)}{\partial x}\right)$$

or, using the more economical notation, as f_{yx} and f_{xy}. Notice that the order of x and y changed in going from the first notational style to the second. In the more detailed style, the partials are taken, and read, from right to left. Thus $\partial^2 f(x, y)/\partial x \partial y$ is calculated by first calculating the partial with respect to y and then calculating the partial with respect to x on the result. In the more economical style, this partial, f_{yx}, is read from left to right. As a practical matter this distinction may not make much of a difference. If $f(x, y)$, f_x, f_y, f_{xy}, and f_{yx} are continuous at some point, then $f_{xy} = f_{yx}$ at that point. (*Note:* Mixed partials are just a special type of second or higher order partial.)

Example: Find both first order partials, both second order partials, and both mixed partials of $f(x, y) = x^3 + 4x^2y - 7xy^2 + 2y^3 + 6$:

$$f_x = 3x^2 + 8xy - 7y^2 \qquad f_y = 4x^2 - 14xy + 6y^2$$
$$f_{xx} = 6x + 8y \qquad f_{yy} = -14x + 12y$$
$$f_{xy} = 8x - 14y \qquad f_{yx} = 8x - 14y$$

In taking the first order partial with respect to x, f_x, all occurrences of the variable y are treated as constants. Thus, when we take the partial with respect to x of the second term, $4x^2y$, we can think of $4y$ as a constant and pull it through the differentiation; that is,

$$\frac{d}{dx}(4x^2y) = 4y \times \frac{d}{dx}x^2 = 4y \times 2x = 8xy.$$

Notice that the last two terms of the equation disappear when we take f_x. This should not be surprising for the last term, which is just the constant 6, because the first rule of differentiation presented in Chapter 3 was that the derivative of a constant is equal to zero. It may seem a bit odd at first that $d(2y^3)/dx = 0$, but this is because y is treated just as any other constant.

Similarly, in calculating f_y the variable x is considered a constant. The first term, x^3, disappears, and other occurrences of the variable x are treated as constants and can be pulled through the differentiation.

In taking the second order partials, begin with one of the first order partials, say f_x, and take the partial derivative with respect to x to obtain f_{xx} and take the partial derivative with respect to y to obtain the mixed partial f_{xy}. In taking a second order partial with respect to x, you treat all occurrences of y (and any other variables present) as constants, regardless of whether you are starting with f_x or f_y. Finally, notice that the two mixed partials are equal to each other, as expected.

Example: Find $f_x, f_y, f_{xx}, f_{yy}, f_{xy}$, and f_{yx} for $f(x, y) = 4x^{-3} - 6x^{2/3}y^{-3} + 8y^5$:

$$f_x = -12x^{-4} - 4x^{-1/3}y^{-3} \qquad f_y = 18x^{2/3}y^{-4} + 40y^4$$

$$f_{xx} = 48x^{-5} + 4/3x^{-4/3}y^{-3} \qquad f_{yy} = -72x^{2/3}y^{-5} + 160y^3$$

$$f_{xy} = 12x^{-1/3}y^{-4} \qquad f_{yx} = 12x^{-1/3}y^{-4}$$

This example has negative and fractional exponents. Nevertheless, the rules are applied as before. As in the previous example, begin by taking the first order partial derivatives, then take the second order partial derivatives from them. Once again, we see that $f_{xy} = f_{yx}$. Given this equality, it would be more economical to just calculate one of these mixed partials knowing the other is equal to it. I suggest, however, that until you are comfortable with calculating derivatives you continue to calculate each partial. If you obtain the same result you can be fairly certain that you took the first and second order partials correctly.

4.2 Extrema of Multivariate Functions

Earlier we saw how to solve max-min (optimization) problems when $y = f(x)$. Now we will extend the method to problems involving more than one independent variable.

Suppose we want to find the high and low points on a smooth surface represented by $z = f(x, y)$, where $f(x, y)$ is defined, is continuous, and has continuous partial derivatives with respect to x and y in some region R in the xy-plane. If there is a point (a, b) in R such that $f(x, y) \geq f(a, b)$ for all points (x, y) sufficiently near to the point (a, b), then the function $f(x, y)$ is said to have a local minimum at (a, b). If the inequality holds for all points

(x, y) in R then $f(x, y)$ has an absolute minimum over R at (a, b). If $f(x, y) \leq f(a, b)$, then (a, b) is a maximum (local or absolute).

If we want to find extrema of $f(x, y)$ over some region R, then the *first (necessary) condition* is to find the set of points that satisfy $f_x = 0$ and $f_y = 0$. In other words, for a point to be an extremum, the slope of the line tangent to the surface in the xz-plane must be equal to zero while holding y constant, and the slope of the line tangent to the surface in the yz-plane must be equal to zero while holding x constant. If only one point satisfies $f_x = f_y = 0$ then it must be the absolute maximum or minimum (assuming it is not a point of inflection or a saddle point). If more than one point satisfies the equality then the *second (sufficient) condition* for one of the points, say $f(a, b)$, to be an absolute maximum (or minimum) is that it be larger (or smaller) than any of the other relative extrema.

Note: Some texts may say "a maximum of f occurs at the point (a, b)." As we saw in Chapter 1 a point is usually designated by a series of values enclosed by parentheses, one value for each dimension. For $y = f(x)$ we refer to points as (x, y). By extension, for $z = f(x, y)$ we would refer to points in the three-dimensional space using three values such as (x, y, z) or (a, b, c) if we are talking about a specific point. One must be careful, therefore, to understand whether an author is speaking of only the dimensions associated with the independent variables.

Example: Find the high or low point of $z = x^2 + 3xy + 3y^2 - 6x + 3y - 6$. Before beginning the calculations, notice that there are two squared terms, one for x and one for y. These two squared terms will cause z to get bigger and bigger as x and y take on larger and larger positive or negative values, despite the xy term. Thus this function is likely to have an absolute minimum but no absolute maximum. Begin by calculating the first order partial derivatives:

$$f_x = 2x + 3y - 6 \quad \text{and} \quad f_y = 3x + 6y + 3$$

Then set both partials equal to zero and solve the system of equations. There are several approaches one can take to solve a system of equations such as this. Some methods will be described in detail in Chapter 6, but for now, begin by setting the two equations equal to each other and manipulating the terms:

$$2x + 3y - 6 = 3x + 6y + 3 \Rightarrow 0 = x + 3y + 9 \Rightarrow -x - 9 = 3y$$

After setting the two functions equal to each other, subtract $2x + 3y - 6$ from both sides of the equation. One of the remaining terms is $3y$, so subtract $x + 9$ from both sides. Now we see what $3y$ is equal to in terms of x. This is important because this value can be substituted for the $3y$ term that appears in f_x. That will leave one equation with only one unknown, which is easily solved:

$$\begin{aligned}0 &= 2x + 3y - 6 = 2x + (-x - 9) - 6\\ &= 2x - x - 9 - 6 = x - 15\\ x &= 15\end{aligned}$$

Now we know that the x-coordinate for the critical point is 15. Next find the value of the y-coordinate by substituting 15 for x in f_x:

$$\begin{aligned}0 &= 2x + 3y - 6 = 2 \times 15 + 3y - 6\\ &= 30 + 3y - 6 = 3y + 24\\ 3y &= -24\\ y &= -8\end{aligned}$$

We could have used f_y to obtain the y-coordinate value, but by using f_x, we can now use f_y to check the x and y values:

$$0 = 3x + 6y + 3 = 3 \times 15 + 6 \times -8 + 3 = 45 - 48 + 3$$

Because the x and y values also satisfy $f_y = 0$, next enter these values into $f(x, y)$ to determine the z-coordinate of the critical point:

$$\begin{aligned}f(x, y) &= x^2 + 3xy + 3y^2 - 6x + 3y - 6\\ f(15, -8) &= (15)^2 + 3(15)(-8) + 3(-8)^2 - 6(15) + 3(-8) - 6\\ &= 225 - 360 + 192 - 90 - 24 - 6\\ &= -63\end{aligned}$$

Thus the critical point is (15, –8, –63). All that remains is to verify that this point is a minimum (according to our expectations). One way to do this is to simply enter values for x and y near the critical point and calculate the z value to see whether it is greater than –63. Unfortunately, this method is not very "scientific." A more rigorous approach is to check the behavior of the difference, $D = f(15 + h, -8 + k) - f(15, -8)$, from the critical point, where h and k are small differences from the x and y values. Entering these values into the original function and simplifying we find:

$$D = h^2 + 3hk + 3k^2 = h^2 + 3hk + \frac{9k^2}{4} + \frac{3k^2}{4} = \left(h + \frac{3k}{2}\right)^2 + \frac{3k^2}{4}$$

We can see that regardless of the value of h and k, the value of D will always be positive because each term on the right side of the equation is squared. Thus (15, –8, –63) is an absolute minimum of $f(x, y)$.

You may wonder how you were to have known that you should manipulate the right side of the equation to get it into a form where it would be the sum of two squared terms. As I tell my students, you weren't. This is the kind of mathematical technique ("trick") that must be seen at least once before it can be applied in other contexts. Although the technique is useful, it can be tedious if the function is complex.

The *second derivative test* is more straightforward and uses the second order partial derivatives of the function. Suppose that (a, b) is a critical point of $f(x, y)$ such that $f_x(a, b) = 0$ and $f_y(a, b) = 0$ (i.e., the first order partial derivatives evaluated at (a, b) equal zero). To see what kind of extremum (a, b) is, if any, we form the *discriminant*:

$$D(a, b) = [f_{xx}(a, b) \times f_{yy}(a, b)] - [f_{xy}(a, b)]^2$$

(*Note:* The last term is squared because we assume $f_{xy} = f_{yx}$.) If $D(a, b) < 0$, the function does not have an extremum at (a, b). If $D(a, b) > 0$ and $f_{xx} < 0$, a local maximum occurs at (a, b). If $D(a, b) > 0$ and $f_{xx} > 0$, a local minimum occurs at (a, b). If $D(a, b) = 0$, the test gives no information about (a, b).

Example: Find all local extrema of $f(x, y) = x^2 + 3xy + 3y^2 - 6x + 3y - 6$ using the second derivative test. This is the same function as in the previous example. Recall that the critical point is (15, –8, –63). Now calculate the second order and mixed partial derivatives of $f(x, y)$:

$$f_x = 2x + 3y - 6 \qquad f_y = 3x + 6y + 3$$

$$f_{xx} = 2 \qquad f_{yy} = 6$$

$$f_{xy} = 3 \qquad f_{yx} = 3$$

Now form a discriminant to test whether this critical point is a minimum:

$$\begin{aligned} D(15,-8) &= [f_{xx}(15,-8) \times f_{yy}(15,-8)] - [f_{xy}(15,-8)]^2 \\ &= (2 \times 6) - (3)^2 = 12 - 9 \\ &= 3 \end{aligned}$$

Thus $D(15, -8) = 3 > 0$ and $f_{xx} = 2 > 0$ so there is a minimum at (15, –8, –63). Because this is the only critical point, and given what we know of the function, this must be an absolute minimum. (*Note:* A more generalized test for problems in higher dimensions, i.e., more independent variables, will be presented in Chapter 6.)

4.2.1 The Method of Least Squares

I have previously mentioned the importance of regression analysis to the social sciences. The essence of regression is to find the line that best fits several data points. The primary method used to find this line is called "least squares" and is a minimization problem. You will most likely use computers to do the calculations, particularly when several independent variables and hundreds or thousands of observations are involved, but you will have a better understanding of your results if you have seen the calculations for a simple bivariate case.

Again, the goal is to find the straight line, $y = mx + b$, that best fits a set of experimental data points, $(x_1, y_1), (x_2, y_2), \ldots (x_n, y_n)$. Associated with each observed x value are two y values: one observed and the other predicted from the equation of the line (i.e., $mx_{\text{obs}} + b$, where x_{obs} is an observed value of x). We shall call the difference between the observed and predicted values of y a *deviation*: $d = y_{\text{obs}} - (mx_{\text{obs}} + b)$. The set of deviations, $d_1 = y_1 - (mx_1 + b)$, $d_2 = y_2 - (mx_2 + b), \ldots, d_n = y_n - (mx_n + b)$, gives an indication of how close the predicted line fits the observed data. It is unlikely that any straight line will be a perfect fit so we must choose a line that in some sense fits "best." (*Note:* Social scientists usually refer to the predicted value of y as "y-hat," denoted $\hat{y}$. This notation is also used in other contexts. Sometimes, however, the difference between observed and predicted or estimated values is distinguished by a change in the type of letter used. For example, in regression it is common to use β_i to refer to the unknown theoretical coefficients of the independent variables, while using b_i to represent the estimated coefficients.)

The method of least squares says that the line, $y = mx + b$, that fits best is the one in which the sum of the squares of the deviations is a minimum. In other words, we want to find the minimum of the function $f(m, b) = d_1^2 +$

$d_2^2 + \ldots + d_n^2$. To do so we solve $f_m = 0$ and $f_b = 0$ simultaneously. (*Note:* The function is in m and b because we *know* the values for x and y and are attempting to determine the values for m and b.)

Example: Find the straight line that best fits the points (0, 1), (1, 3), (2, 2), (3, 4), and (4, 5) according to the method of least squares. Begin by determining the values of the deviations and their squares.

x_{obs}	y_{obs}	$d = y_{obs} - (mx_{obs} + b)$	d^2
0	1	$1 - b$	$1 - 2b + b^2$
1	3	$3 - m - b$	$9 - 6b + b^2 - 6m + 2mb + m^2$
2	2	$2 - 2m - b$	$4 - 4b + b^2 - 8m + 4mb + 4m^2$
3	4	$4 - 3m - b$	$16 - 8b + b^2 - 24m + 6mb + 9m^2$
4	5	$5 - 4m - b$	$25 - 10b + b^2 - 40m + 8mb + 16m^2$

Summing the squares of the deviations yields:

$$f(m, b) = 55 - 30b + 5b^2 - 78m + 20mb + 30m^2$$

Next calculate the first order partial derivatives of $f(m, b)$:

$$f_m = -78 + 20b + 60m \quad \text{and} \quad f_b = -30 + 10b + 20m$$

Now set each partial derivative equal to zero and solve simultaneously. Begin by manipulating f_b as follows:

$$\begin{aligned} 0 &= -30 + 10b + 20m \\ 10b &= 30 - 20m \end{aligned}$$

Now substitute this value into f_m to obtain m:

$$\begin{aligned} 0 &= -78 + 20b + 60m = -78 + 2(10b) + 60m \\ &= -78 + 2(30 - 20m) + 60m = -78 + 60 - 40m + 60m \\ &= -18 + 20m \end{aligned}$$

$$20m = 18$$
$$m = 0.9$$

Substitute this value back into f_b to obtain b:

$$10b = 30 - 20(0.9) = 30 - 18 = 12$$
$$b = 1.2$$

Entering these values into $f(m, b)$, we find that the critical point is (0.9, 1.2, 1.9). To verify that (0.9, 1.2, 1.9) is a minimum, look at the differences around it:

$$f(0.9 + h, 1.2 + k) - f(0.9, 1.2) = 5k^2 + 20hk + 30h^2 = 5(k + 2h)^2 + 10h^2$$

The difference is greater than zero for all nonzero h and k so it is a minimum. For the bivariate case, this is known as Ordinary Least Squares (OLS) regression.

4.2.2 Constrained Optimization (Method of Lagrange Multipliers)

Suppose that we want to maximize (or minimize) a function $f(x, y)$ subject to some constraint $g(x, y) = 0$. (Perhaps we have a production function that is subject to a budgetary constraint or perhaps we would like to pay for a new government program whose spending is limited to a new consumption tax.) Begin by creating a new function composed of $f(x, y)$ and $g(x, y)$ as follows:

$$F(x, y, \lambda) = f(x, y) - \lambda \times g(x, y)$$

where λ is a new variable multiplied by $g(x, y)$. Notice that $F(x, y, \lambda) = f(x, y)$ because $g(x, y) = 0$. Then why make $F(x, y, \lambda)$? The use of $g(x, y)$ allows us to limit the values of x and y under consideration without changing the values of $f(x, y)$. More specifically, by adding λ and maximizing with respect to it and the other variables, we limit our search for extrema to those values of x and y that satisfy $g(x, y) = 0$. Just as with previous optimization problems, the next step is to set each of the first order partial derivatives equal to zero and solve simultaneously: $0 = F_x = F_y = F_\lambda$.

Of course, F_λ is the first partial derivative of $F(x, y, \lambda)$ with respect to λ. Each of these partials can be rewritten in terms of $f(x, y)$ and $g(x, y)$ as follows:

$$0 = F_x = f_x - \lambda \times g_x$$
$$0 = F_y = f_y - \lambda \times g_y$$
$$0 = F_\lambda = -g(x, y)$$

Because $f(x, y)$ does not contain λ, it is treated as a constant when calculating F_λ. This means $F_\lambda = 0$ because $g(x, y) = 0$, and thus $-g(x, y) = 0$. After finding the three first order partial derivatives, we are left with three equations and three unknowns, with the constraint as one of the equations. If the equations are simple, you can manipulate them to isolate one of the variables, solve for it, then substitute back into one of the other equations to solve for a second variable, then substitute both into the third equation to solve for the remaining variable. The method of Lagrange multipliers also works for functions and constraints with more variables, provided you have more variables than constraints. (This function is in two variables and there is one constraint.) Of course, the calculations will usually be done by a computer program.

Example: Using Lagrange multipliers find the extrema of $f(x, y) = xy$ that lie on the unit circle $x^2 + y^2 = 1$. First recognize that $g(x, y) = x^2 + y^2 - 1 = 0$ is the constraint. Next construct $F(x, y, \lambda)$:

$$F(x, y, \lambda) = f(x, y) - \lambda \times g(x, y) = xy - \lambda(x^2 + y^2 - 1)$$

Now calculate the first order partial derivatives of $F(x, y, \lambda)$ and set them equal to zero:

$$F_x = y - 2\lambda x = 0$$
$$F_y = x - 2\lambda y = 0$$
$$F_\lambda = -(x^2 + y^2 - 1) = 0$$

As previously noted, there is no set approach to solving such a system of equations. For this system, start by isolating y in F_x and substituting it into F_y:

$$0 = y - 2\lambda x$$
$$y = 2\lambda x$$
$$0 = x - 2\lambda y = x - 2\lambda(2\lambda x) = x - 4\lambda^2 x$$

$$x = 4\lambda^2 x$$
$$1 = 4\lambda^2$$
$$\lambda^2 = 1/4$$
$$\lambda = \pm 1/2$$

Now take these values for λ and enter them into F_x:

$$y - 2\lambda x = 0$$
$$y - 2(-1/2)x = 0 \Rightarrow y + x = 0 \Rightarrow y = -x$$
$$y - 2(1/2)x = 0 \Rightarrow y - x = 0 \Rightarrow y = x$$

Next substitute these values for y into F_λ:

$$x^2 + y^2 - 1 = 0$$
$$x^2 + (-x)^2 - 1 = 0 \Rightarrow 2x^2 = 1 \Rightarrow x^2 = 1/2 \Rightarrow x = \pm 1/\sqrt{2}$$
$$x^2 + x^2 - 1 = 0 \Rightarrow 2x^2 = 1 \Rightarrow x^2 = 1/2 \Rightarrow x = \pm 1/\sqrt{2}$$

Now substitute these two values for x back into F_λ to determine that $y = \pm 1/\sqrt{2}$. Entering these values into $f(x, y)$ yields four critical points: $(1/\sqrt{2}, 1/\sqrt{2}, 1/2)$, $(-1/\sqrt{2}, -1/\sqrt{2}, 1/2)$, $(-1/\sqrt{2}, 1/\sqrt{2}, -1/2)$, and $(1/\sqrt{2}, -1/\sqrt{2}, -1/2)$. Verify for yourself that the first two points are maxima and the second two are minima. (*Note:* If you check these points by entering other values for x and y, do not forget that they must satisfy the constraint.)

We will use Lagrange multipliers and constrained optimization again in Chapter 6.

5. INTEGRAL CALCULUS

Integration, also known as antidifferentiation, is the inverse of differentiation. Recall that if we had a function that indicates Carl Lewis's distance from the starting line, then the derivative of that function yields his instantaneous speed. Suppose, however, that we begin with a function for Lewis's instantaneous speed. We could then *integrate* this function to obtain his distance from the starting line. A primary use of integration for social scientists is to determine the area under a curve. In most instances, the curve is a probability curve and the researcher wants to determine the cumulative probability that some event will occur given certain parameters.

We will discuss two types of integrals, definite and indefinite. A function F is the *indefinite integral* of a function f on an interval I iff $F'(x) = f(x)$ for all $x \in I$. In other words, F is the indefinite integral of another function f if you obtain f when you take the derivative of F. Recall that when we take the derivative of a function any constant terms disappear. If we do not know what the original function was and try to go back to it via integration, we will not know what the constant term, if any, was. Thus indefinite integrals are not unique. If F is an indefinite integral of f and c is a constant, then $G(x) = F(x) + c$ is a family of functions that differ only by the size of the constant term; that is, they are shifted up or down the y-axis. The usual notation for indefinite integrals is $\int f(x)\,dx = F(x) + c$. The symbol $\int$ is called the *integral sign*. In the above expression, $f(x)$, the function you are integrating, is called the *integrand*. The constant in the expression, c, is called the *constant of integration*. The dx in the expression is primarily to indicate that the integration is taking place with respect to a particular variable, in this case x. If the variable of interest was t, then you would use dt as the indicator. Notice that $\int$ and dx enclose the integrand. This means that you integrate everything between them.

5.1 Integration Rules

Because integration is the reverse (or inverse) of differentiation, many of the rules for integration can be obtained by using the rules of differentiation in reverse.

1. $\int x^n\,dx = \dfrac{x^{n+1}}{n+1} + c$: This is known as the power rule for integration.

   Notice that if you take the derivative of $\dfrac{x^{n+1}}{n+1} + c$ you get x^n, the original function.

2. $\int x^{-1}\,dx = \ln|x| + c$: This rule fills the gap in the power rule for x^{-1} (which would not work because there would be a 0 in the denominator). Notice that we take the natural log of the *absolute value* of x. This is necessary because we cannot take the log, natural or otherwise, of a negative number.

3. $\int dx = x + c$: This is actually a special case of the power rule because $dx = 1\ dx = x^0\ dx$. Applying the power rule to x^0 yields $\frac{x^{0+1}}{0+1} + c = x + c$.
4. $\int [a \times f(x)]\ dx = a \int f(x)\ dx$: The integral of a constant times a function is equal to the constant times the integral of the function. In other words, you can pull a constant through the integration.
5. $\int [f(x) + g(x)]\ dx = \int f(x)\ dx + \int g(x)\ dx$: The integral of the sum of two functions is equal to the sum of their integrals.
6. $\int f[g(x)] \times g'(x)\ dx = F[g(x)] + c$: This is the chain rule for integration. The integral of a function f evaluated at another function g and multiplied by the derivative of g is equal to F evaluated at g plus the constant of integration. This rule is often difficult to grasp at first and the examples below give a detailed explanation of how the rule works.
7. $\int [g(x) \times h(x)]\ dx = g(x) \times H(x) - \int [g'(x) \times H(x)]\ dx + c$, where $H(x) = \int h(x)\ dx$: This is the formula for the technique known as *integration by parts*. The integral of two functions is equal to the first function times the antiderivative of the second function minus the integral of the derivative of the first function times the antiderivative of the second function plus the constant of integration.
8. $\int e^x\ dx = e^x + c$; $\int a^x\ dx = \frac{a^x}{\ln a} + c$: These rules show the reverse of differentiation rule 9 from Chapter 3.

You may have noticed that there are no rules for products or quotients (except for certain special cases in rules 6 and 7). This is because no general rules for these operations exist. Integration is usually much more difficult than differentiation and often involves working from tables of integrals for known functions. Sometimes the difficulty is in getting the integrand into a form that you know how to integrate. (This is usually the problem in applying rule 6.) Fortunately, most functions used by social scientists are either relatively easy to integrate or have known integrals.

Examples: Find the indefinite integrals for the following functions.

1. $\int (4x+3)\,dx$: When you first start doing integration you should break down the problem as much as possible. When you feel more comfortable using the rules you can begin to "skip steps."

$$\int (4x+3)\,dx = \int 4x\,dx + \int 3\,dx = 4\int x\,dx + 3\int dx$$
$$= 4 \times \frac{x^{1+1}}{1+1} + c_1 + 3x + c_2 = 2x^2 + 3x + c_3$$

Rules 1, 3, 4, and 5 were used to break this integral into much easier integrals. Notice the subscripts on the constants of integration. Breaking the original integral into two easier ones yielded two constants of integration instead of just one. In the final line these two constants were added together to form a third constant ($c_1 + c_2 = c_3$). We do not know the value of these constants, so we need not pay too much attention to them. Nevertheless, you should distinguish between them so you remember that they have different values. Remember that you can check to see if you obtained the correct integral by taking its derivative. If you get the original function back, then you did the integration correctly.

2. $\int \left(\frac{1}{x^3} + \frac{1}{\sqrt[3]{x}}\right) dx$: This example may appear difficult, but it is really a straightforward application of rules 1 and 5. Begin by putting the integrand into a form that is more recognizable:

$$\int \left(\frac{1}{x^3} + \frac{1}{\sqrt[3]{x}}\right) dx = \int (x^{-3} + x^{-1/3})\,dx = \int x^{-3}\,dx + \int x^{-1/3}\,dx$$
$$= \frac{x^{-3+1}}{-3+1} + c_1 + \frac{x^{-1/3+1}}{-1/3+1} + c_2 = -\frac{x^{-2}}{2} + \frac{x^{2/3}}{2/3} + c_3$$
$$= -\frac{1}{2x^2} + \frac{3x^{2/3}}{2} + c_3$$

How you leave the answer is a matter of preference, but you should try to leave the answer in an easily readable form.

3. $\int (4x+3)^{1/3}\,dx$: Here we will apply the chain rule for integration. Use of the chain rule involves a technique known as *substitution*. Let $u =$ $4x + 3$. Then $\frac{du}{dx} = 4 \Rightarrow du = 4\,dx \Rightarrow dx = \frac{du}{4}$. Now make these substitutions:

$$\int (4x+3)^{1/3}\,dx = \int u^{1/3}\frac{du}{4}$$

Pull the constant ¼ through the integration and use the power rule:

$$\int u^{1/3}\frac{du}{4} = \frac{1}{4}\int u^{1/3}\,du = \frac{1}{4}\left(\frac{u^{4/3}}{4/3} + c_1\right) = \frac{3u^{4/3}}{16} + c_2$$

Reverse the substitutions to obtain an answer in terms of x:

$$\frac{3u^{4/3}}{16} + c_2 = \frac{3(4x+3)^{4/3}}{16} + c_2$$

Caution: This method worked because we treated du and dx as quantities that could be manipulated just as any other variable or number. Though I will not explain why here, this is not an entirely accurate interpretation of du and dx, but it is acceptable for making substitutions for purposes of integration.

4. $\int t(4t^2+2)^2\,dt$: Although the variable of interest here is t, it does not affect the application of the rules of integration. (If, however, dx was still listed, then the variable of interest would still be x and all appearances of t would be treated as constants.) As in the previous example, we begin by letting $u = 4t^2 + 2$, then $\frac{du}{dt} = 8t \Rightarrow du = 8t\,dt$ $\Rightarrow \frac{du}{8} = t\,dt$. Although you could not tell from the previous example, the purpose of these manipulations is to isolate each variable on opposite sides of the equal sign. Thus here we leave the t with the dt. Fortunately, the t at the front of the integrand can be grouped with the dt so the substitution can be made:

$$\int t(4t^2+2)^2\,dt = \int (4t^2+2)^2 t\,dt = \int u^2\frac{du}{8}$$

Now do the integration,

$$\int u^2 \frac{du}{8} = \frac{1}{8}\int u^2\, du = \frac{1}{8}\left(\frac{u^3}{3} + c_1\right) = \frac{u^3}{24} + c_2$$

and make the substitution back to the variable t:

$$\frac{u^3}{24} + c_2 = \frac{(4t^2+2)^3}{24} + c_2$$

To do this integration we needed to have an additional factor of t to the same power as the derivative of the function inside the parentheses. Here the power of t inside the parentheses was two, so there had to be a t to the first power outside the parentheses to do the integration.

5. $\int \frac{z^3}{z^4+3}\, dz$: Let $u = z^4 + 3$, then $\frac{du}{dz} = 4z^3 \rightarrow du = 4z^3\, dz \rightarrow \frac{du}{4} = z^3\, dz$.

Because we have a factor of z^3, we can make the substitutions and do the integration:

$$\int \frac{z^3}{z^4+3}\, dz = \int \frac{1}{u}\frac{du}{4} = \frac{1}{4}\int u^{-1}\, du$$

$$= \frac{1}{4}(\ln|u| + c_1) = \frac{\ln|u|}{4} + c_2$$

Now make the substitution back to z:

$$\frac{\ln|u|}{4} + c_2 = \frac{\ln|z^4+3|}{4} + c_2$$

6. $\int \frac{e^{\sqrt[3]{x}}}{\sqrt[3]{x^2}}\, dx$: This example looks formidable in its present form. Begin by converting all radical signs ($\sqrt{\ }$) to exponents. After doing so we have

$$\int \frac{e^{\sqrt[3]{x}}}{\sqrt[3]{x^2}}\, dx = \int \frac{e^{x^{1/3}}}{x^{2/3}}\, dx = \int e^{x^{1/3}} x^{-2/3}\, dx$$

I took the additional step of moving all the factors to the numerator. Now let $u = x^{1/3}$, then $\frac{du}{dx} = \frac{1}{3}x^{-2/3} \Rightarrow du = \frac{1}{3}x^{-2/3}\,dx \Rightarrow 3du = x^{-2/3}\,dx$. After making the substitutions and doing the integration we find:

$$\int e^{x^{1/3}} x^{-2/3}\,dx = \int e^u \times 3\,du = 3\int e^u\,du$$
$$= 3(e^u + c_1) = 3e^u + c_2$$

Substituting back to x yields:

$$3e^u + c_2 = 3e^{x^{1/3}} + c_2$$

7. $\int x \ln x\,dx$: Here we will use integration by parts. Let $g(x) = \ln x$ and $h(x) = x$. Then $H(x) = \frac{x^2}{2} + c_1$ and $g'(x) = \frac{1}{x}$. Now enter these values into the formula:

$$\int [g(x) \times h(x)]\,dx = g(x) \times H(x) - \int [g'(x) \times H(x)]\,dx + c$$
$$= (\ln x)\left(\frac{x^2}{2} + c_1\right) - \int \left(\frac{1}{x}\right)\left(\frac{x^2}{2} + c_1\right)dx$$
$$= \frac{x^2 \ln x}{2} + c_1 \ln x - \int \left(\frac{x}{2} + \frac{c_1}{x}\right)dx$$

Although we are left with another integral, this one is much easier to evaluate than the first one. Now continue with the second integration:

$$= \frac{x^2 \ln x}{2} + c_1 \ln x - \int \left(\frac{x}{2} + \frac{c_1}{x}\right)dx = \frac{x^2 \ln x}{2} + c_1 \ln x - \left(\frac{x^2}{4} + c_1 \ln x + c_2\right)$$

$$= \frac{x^2 \ln x}{2} + c_1 \ln x - \frac{x^2}{4} - c_1 \ln x - c_2 = \frac{x^2 \ln x}{2} - \frac{x^2}{4} - c_2 = \left(\frac{x^2}{2}\right)\left(\ln x - \frac{1}{2}\right) - c_2$$

Note that the first constant of integration, c_1, drops out and we are just left with c_2. (The negative sign on c_2 is not really necessary because we do not know what c_2 is, but keeping it is more precise given the calculations.)

8. $\int xe^{2x}\,dx$: Let $g(x) = x$ and $h(x) = e^{2x}$. Then $g'(x) = 1$ and $H(x) = \frac{e^{2x}}{2} + c_1$. Enter these values into the formula and do the integration:

$$g(x) \times H(x) - \int [g'(x) \times H(x)]\,dx + c$$

$$= x\left(\frac{e^{2x}}{2} + c_1\right) - \int 1\left(\frac{e^{2x}}{2} + c_1\right)dx = \frac{xe^{2x}}{2} + c_1 x - \left(\frac{e^{2x}}{4} + c_1 x + c_2\right)$$

$$= \frac{xe^{2x}}{2} + c_1 x - \frac{e^{2x}}{4} - c_1 x - c_2 = \frac{xe^{2x}}{2} - \frac{e^{2x}}{4} - c_2 = \frac{e^{2x}}{2}\left(x - \frac{1}{2}\right) - c_2$$

In performing integration by parts, $g(x)$ and $h(x)$ must be chosen carefully. You must be able to easily differentiate $g(x)$ and integrate $h(x)$. In addition, and more important, the integral $\int [g'(x) \times H(x)]\,dx$ must be easier to evaluate than the original integral. (It is possible that you will need to perform integration by parts a second time on $\int [g'(x) \times H(x)]\,dx$.)

Note: A more common statement of the formula for integration by parts is $\int u\,dv = uv - \int v\,du + c$, where u and v are functions in x, du and dv are the derivatives of u and v, and c is the constant of integration. For those first learning integration by parts, this version of the formula seems confusing because one must view the original integral as a product of a function and a derivative of a function rather than the product of two functions. The statement of the formula in rule 7 is more straightforward and is also consistent with previous usage (in particular the product rule—see Chapter 3).

5.2 The Theory of Integration and Definite Integrals

Suppose we wish to determine the area under a curve on an interval $[x_1, x_3]$. We can get an approximation of the area using Darboux's Theorem. Consider the function graphed in Figure 5.1a. We could estimate the area under the curve between x_1 and x_3 to be $y_1(x_3 - x_1)$, which is the height times the width of the small rectangle under the curve, but this would severely underestimate the area. We could also estimate the area to be $y_3(x_3 - x_1)$,

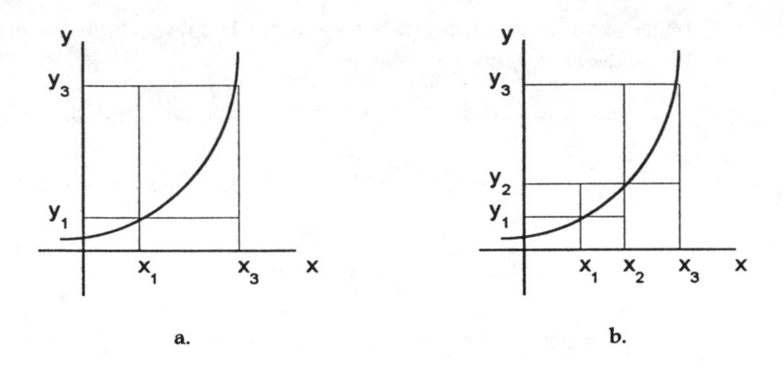

Figure 5.1 Estimating the Area Under a Curve

but this would overestimate the area. We would get a better approximation if we divided the interval into two equal parts and estimated the area again. If x_2 is the midpoint of $[x_1, x_3]$, then we could estimate the area to be $y_1(x_2 - x_1) + y_2(x_3 - x_2)$, which we can call the lower sum (see Figure 5.1b). This is a better estimate than $y_1(x_3 - x_1)$, but it still underestimates the area. Similarly, we could estimate the area to be $y_2(x_2 - x_1) + y_3(x_3 - x_2)$, which we can call the upper sum. Again, this is a better estimate than $y_3(x_3 - x_1)$, but it still overestimates the area.

As you may have guessed, if we continue to divide the interval into smaller parts our estimates get better and, incidentally, the upper and lower sums get closer to each other. According to Darboux's Theorem, we can form the lower Darboux sum $\sum_{i=1}^{n} m_i(x_i - x_{i-1})$ and the upper Darboux sum $\sum_{i=1}^{n} M_i(x_i - x_{i-1})$, where m_i is the greatest lower bound of $f(x)$ on the interval $[x_{i-1}, x_i]$ (the smaller rectangles) and M_i is the least upper bound on the same interval (the larger rectangles). The limits of these sums as $n \to \infty$ (or as the length of the subinterval goes to 0) give two integrals known as the lower Darboux integral (for lower bounds) and the upper Darboux integral (for upper bounds). If both limits exist and are equal to each other, the function f is said to be integrable on the interval.

This leads us to the *fundamental theorem of calculus*: If $\int f(x)\,dx$ exists and a function $F(x)$ exists such that $F'(x) = f(x)$ for all $x \in [a, b]$ then

$$\int_a^b f(x)\,dx = F(x)\,\Big|_a^b = F(b) - F(a)$$

This formula means that the area under a curve is equal to the integral of the function evaluated at the right endpoint of the interval minus the integral evaluated at the left endpoint. (*Note:* Direction makes a difference. If you switch the endpoints you change the sign of the area.) The constants of integration cancel out so we do not write them.

Here are a few rules and results regarding definite integrals:

1. If a function is continuous on $[a, b]$, then it is integrable on $[a, b]$.
2. $\int_a^b f(x)\,dx = -\int_b^a f(x)\,dx$: As noted, changing the direction changes the sign.
3. $\int_a^a f(x)\,dx = 0$
4. $\int_a^b dx = b - a$: This is the area under the constant function $f(x) = 1$.
5. $\int_a^b [f(x) + g(x)]\,dx = \int_a^b f(x)\,dx + \int_a^b g(x)\,dx$: This tells us that rule 5 for indefinite integrals applies to definite integrals as well.
6. For $a < c < b$, $\int_a^b f(x)\,dx = \int_a^c f(x)\,dx + \int_c^b f(x)\,dx$: The area under a curve for an interval is equal to the area under the curve for the sum of the subintervals. (This rule is also true for $a \le c \le b$ because of rule 3.)
7. For a function f, the average value of f on $[a, b]$ is $\dfrac{\int_a^b f(x)\,dx}{b - a}$

Examples: Find the area under the curve for the indicated interval for each of these definite integrals.

1. $\int_1^3 (x^2 + 1)\,dx$:

$$\int_1^3 (x^2+1)\,dx = \int_1^3 x^2\,dx + \int_1^3 1\,dx = \left(\frac{x^3}{3}+c_1\right)\Bigg|_1^3 + (x+c_2)\,\Big|_1^3 = \left(\frac{x^3}{3}+x+c_3\right)\Bigg|_1^3$$

The first step was to break the integral into two parts using rule 5. The second step was to do the integration, but I did not calculate the area under the curve for the interval [1,3]. This is what the vertical line to the right of each set of parentheses indicates. As you can see, the value at the top of the vertical line corresponds to the value at the top of the integral sign and the value at the bottom of the vertical line corresponds to the value at the bottom of the integral sign. The length of the vertical line varies only because of the size of the function to which it is attached. Next, rule 5 is used again to recombine the two parts. (The constants of integration are included in this example to show how they cancel out. They are not usually written for definite integrals.)

$$\left(\frac{x^3}{3}+x+c_3\right)\Bigg|_1^3 = \left(\frac{3^3}{3}+3+c_3\right)-\left(\frac{1^3}{3}+1+c_3\right)$$
$$= 9+3+c_3-\frac{1}{3}-1-c_3$$
$$= 12 - 4/3 = 10\tfrac{2}{3}$$

2. $\int_1^4 (x^3+4x^2-7x-3)\,dx$: Rule 5 will again be used for this example, but not every step in the calculations will be shown:

$$\int_1^4 (x^3+4x^2-7x-3)\,dx = \left(\frac{x^4}{4}+\frac{4x^3}{3}-\frac{7x^2}{2}-3x\right)\Bigg|_1^4$$

$$= \left(64+\frac{256}{3}-56-12\right)-\left(\frac{1}{4}+\frac{4}{3}-\frac{7}{2}-3\right)$$

$$= \frac{244}{3}-\left(-\frac{59}{12}\right)=\frac{1035}{12}=86\tfrac{1}{4}$$

3. $\int_0^2 x^3(x^4+1)^{1/2}\,dx$: This example will use the chain rule for integration.

Let $u = x^4 + 1$, then $\frac{du}{dx} = 4x^3 \Rightarrow du = 4x^3\,dx \Rightarrow \frac{du}{4} = x^3\,dx$. Now make the substitutions and do the integration:

$$\int_0^2 x^3(x^4+1)^{1/2}\,dx = \int_0^2 u^{1/2}\frac{du}{4} = \frac{1}{4}\int_0^2 u^{1/2}\,du$$

$$= \frac{1}{4}\left(\frac{u^{3/2}}{3/2}\right)\Bigg|_0^2 = \left(\frac{u^{3/2}}{6}\right)\Bigg|_0^2$$

Now make the substitution back to x and finish the calculation:

$$\left(\frac{u^{3/2}}{6}\right)\Bigg|_0^2 = \left(\frac{(x^4+1)^{3/2}}{6}\right)\Bigg|_0^2 = \left(\frac{(2^4+1)^{3/2}}{6}\right) - \left(\frac{(0^4+1)^{3/2}}{6}\right)$$

$$= \frac{17^{3/2}-1}{6} \approx 11.52$$

4. $\int_1^2 (x^3+1)\ln x\,dx$: This example will require integration by parts. Begin by calculating the indefinite integral. Let $g(x) = \ln x$ and $h(x) = x^3 + 1$. Then $g'(x) = \frac{1}{x}$ and $H(x) = \frac{x^4}{4} + x + c_1$. Enter these values into the formula:

$$\int [g(x) \times h(x)]\,dx = g(x) \times H(x) - \int [g'(x) \times H(x)]\,dx + c$$

$$= (\ln x)\left(\frac{x^4}{4} + x + c_1\right) - \int \left(\frac{1}{x}\right)\left(\frac{x^4}{4} + x + c_1\right)dx$$

$$= \frac{x^4 \ln x}{4} + x\ln x + c_1 \ln x - \int \left(\frac{x^3}{4} + 1 + \frac{c_1}{x}\right)dx$$

$$= \frac{x^4 \ln x}{4} + x \ln x + c_1 \ln x - \left(\frac{x^4}{16} + x + c_1 \ln x + c_2 \right)$$

$$= \frac{x^4 \ln x}{4} + x \ln x + c_1 \ln x - \frac{x^4}{16} - x - c_1 \ln x - c_2$$

$$= \frac{x^4 \ln x}{4} + x \ln x - \frac{x^4}{16} - x - c_2$$

Rewrite the last step as for a definite integral and finish the calculations:

$$\frac{x^4 \ln x}{4} + x \ln x - \frac{x^4}{16} - x \Big|_1^2 = \frac{2^4 \ln 2}{4} + 2 \ln 2 - \frac{2^4}{16} - 2 - \left(\frac{1^4 \ln 1}{4} + 1 \ln 1 - \frac{1^4}{16} - 1 \right)$$

$$\approx 4(.6931) + 2(.6931) - 3 - \left(\frac{1 \times 0}{4} + (1 \times 0) - \frac{17}{16} \right) \approx 2.2214$$

Social scientists often consider the probabilities associated with the values of continuous variables. We often speak of the probability that an observed value falls within a specified range rather than the probability of a specific outcome. We use probability density functions (pdf) to determine these probabilities: $\Pr(a < x < b) = \int_a^b f(x)\,dx$, where Pr stands for probability and $f(x)$ is the pdf. Probability density functions must satisfy two requirements: (a) $f(x) \geq 0$ for all x (i.e., there can be no negative probabilities) and (b) $\int_{-\infty}^{\infty} f(x)\,dx = 1$ (i.e., the sum of the probabilities for all outcomes *must* equal one). The *mean* of a distribution with pdf f is $E(x) = \mu_x = \int_{-\infty}^{\infty} x \times f(x)\,dx$. The *variance* is $\mathrm{Var}(x) = \int_{-\infty}^{\infty} (x - \mu_x)^2 \times f(x)\,dx$.

Example: Let $f(x)$ be a probability density function, where

$$f(x) = \begin{cases} cx, & 0 \leq x \leq 4 \\ 0, & \text{otherwise} \end{cases}$$

Find c, the mean of $f(x)$, and the variance of $f(x)$. To begin, we know $\int_0^4 cx\,dx = 1$. From this we can determine c as follows:

$$\int_0^4 cx\,dx = 1 \Rightarrow \left(\frac{cx^2}{2} \right)\Big|_0^4 = 1 \Rightarrow 8c = 1 \Rightarrow c = 1/8.$$

We can rewrite the pdf as

$$f(x) = \begin{cases} \dfrac{x}{8}, & 0 \le x \le 4 \\ 0, & \text{otherwise} \end{cases}$$

Now calculate the mean as

$$\mathrm{E}(x) = \int_0^4 \left(x \times \frac{x}{8} \right) dx = \frac{1}{8} \int_0^4 x^2 \, dx = \frac{1}{8} \left(\frac{x^3}{3} \right) \Bigg|_0^4 = \frac{x^3}{24} \Bigg|_0^4 = \frac{64}{24} = \frac{8}{3}$$

and the variance as

$$\mathrm{Var}(x) = \int_0^4 \left((x - 8/3)^2 \times \frac{x}{8} \right) dx = \frac{1}{8} \int_0^4 x(x - 8/3)^2 \, dx = \frac{1}{8} \int_0^4 \left(x^3 - \frac{16x^2}{3} + \frac{64x}{9} \right) dx$$

$$= \frac{1}{8} \left(\frac{x^4}{4} - \frac{16x^3}{9} + \frac{32x^2}{9} \right) \Bigg|_0^4 = \left[\frac{1}{8} \left(64 - \frac{1024}{9} + \frac{512}{9} \right) - 0 \right]$$

$$= \frac{1}{8} \left(\frac{576 - 1024 + 512}{9} \right) = \frac{1}{8} \times \frac{64}{9} = \frac{8}{9}$$

6. MATRIX ALGEBRA

6.1 Matrices

Social scientists often use data arranged in an ordered array of numbers. For example, one might construct a rectangular array in which each column represents a variable, each row represents an observation, and each element in the array is the value of a specified variable and observation. Data arranged in this manner can be manipulated to give us information about the relationships between the variables and the observations. A *matrix* is an ordered array of elements. The elements can be functions, variables, or numbers. As with many other mathematical concepts, matrices can be represented several ways: $[a_{ij}]$, (a_{ij}), $\|a_{ij}\|$, $\boldsymbol{A}_{m \times n}$, and $\boldsymbol{A}$. The first three forms differ only by the way the element a_{ij} is enclosed and is primarily a matter of taste (or typesetting ability). The element a_{ij} is the value of the entry in the ith row and jth column, counting from the upper left corner. The row designator always precedes the column designator. The last two forms represent the matrix with a boldface letter, sometimes with a subscript indicating the dimensions or *order* of the matrix. The $m \times n$ subscript tells us the matrix has m rows and n columns and is therefore called "an m by n matrix." If a matrix has the same number of rows and columns, $\mathbf{A}_{n \times n}$, it is said to be a *square matrix.* An $m \times 1$ matrix is called a *column vector* (or matrix) because it has only one column and a $1 \times n$ matrix is called a *row vector* (or matrix). A 1×1 matrix is called a *scalar* (or number). The row and column numbering for the generic matrix $\mathbf{A}_{m \times n}$ is as follows:

$$\mathbf{A}_{m \times n} = \begin{bmatrix} a_{11} & a_{12} & a_{13} & \cdots & a_{1n} \\ a_{21} & a_{22} & a_{23} & \cdots & a_{2n} \\ \cdot & \cdot & \cdot & \cdot & \cdot \\ \cdot & \cdot & \cdot & \cdot & \cdot \\ \cdot & \cdot & \cdot & \cdot & \cdot \\ a_{m1} & a_{m2} & a_{m3} & \cdots & a_{mn} \end{bmatrix}$$

6.1.1 Matrix Rules

1. Two matrices $\mathbf{A} = [a_{ij}]$ and $\mathbf{B} = [b_{ij}]$ are equal iff $\mathbf{A}$ and $\mathbf{B}$ have the same order (i.e., the same number of rows and columns) and $a_{ij} = b_{ij}$ for all i and j.

2. For $\mathbf{A} = [a_{ij}]$ and $\mathbf{B} = [b_{ij}]$ with equal order, $m \times n$, the matrix $\mathbf{A} + \mathbf{B}$ is the matrix of order $m \times n$ such that $\mathbf{A} + \mathbf{B} = [a_{ij} + b_{ij}]$. In addition, the commutative law ($\mathbf{A} + \mathbf{B} = \mathbf{B} + \mathbf{A}$) and the associative law [($\mathbf{A}$ + $\mathbf{B}$) + $\mathbf{C}$ = $\mathbf{A}$ + ($\mathbf{B}$ + $\mathbf{C}$)] both hold for matrix addition.

Example: If $\mathbf{A} = \begin{bmatrix} 6 & 3 \\ -1 & 2 \end{bmatrix}$ and $\mathbf{B} = \begin{bmatrix} 1 & 4 \\ 5 & -2 \end{bmatrix}$, then

$$\mathbf{A} + \mathbf{B} = \begin{bmatrix} 6+1 & 3+4 \\ -1+5 & 2+(-2) \end{bmatrix} = \begin{bmatrix} 7 & 7 \\ 4 & 0 \end{bmatrix}.$$

3. If $\mathbf{A} = [a_{ij}]$ and k is a scalar (constant), then the matrix $k\mathbf{A} = [ka_{ij}] = \mathbf{A}k$. If you multiply a matrix by a scalar, then *every* element of the

matrix is multiplied by that scalar. Multiplication by a scalar is commutative.

Example: If $\boldsymbol{A} = \begin{bmatrix} 6 & 3 \\ -1 & 2 \end{bmatrix}$ and $k = 3$, then

$$k\boldsymbol{A} = 3\begin{bmatrix} 6 & 3 \\ -1 & 2 \end{bmatrix} = \begin{bmatrix} 3\times 6 & 3\times 3 \\ 3\times -1 & 3\times 2 \end{bmatrix} = \begin{bmatrix} 18 & 9 \\ -3 & 6 \end{bmatrix}.$$

4. Given rules 2 and 3, matrix subtraction can be defined as the addition of two matrices, one of which is multiplied by the scalar -1: $\boldsymbol{A} - \boldsymbol{B} = \boldsymbol{A} + (-1)\boldsymbol{B}$.

Example: If $\boldsymbol{A}$ and $\boldsymbol{B}$ are as above, then

$$\boldsymbol{A} - \boldsymbol{B} = \begin{bmatrix} 6-1 & 3-4 \\ -1-5 & 2-(-2) \end{bmatrix} = \begin{bmatrix} 5 & -1 \\ -6 & 4 \end{bmatrix}.$$

5. Matrix multiplication: If you have two matrices, $\boldsymbol{A} = [a_{ik}]$ of order $m \times p$ and $\boldsymbol{B} = [b_{kj}]$ of order $p \times n$, the matrix $\boldsymbol{AB} = \boldsymbol{C} = [c_{ij}]$ has order $m \times n$ where $c_{ij} = \sum_{k=1}^{p} a_{ik} b_{kj}$. In calculating $\boldsymbol{AB}$, we say that $\boldsymbol{A}$ is *postmultiplied* by $\boldsymbol{B}$ or that $\boldsymbol{B}$ is *premultiplied* by $\boldsymbol{A}$.

Example: If $\boldsymbol{A}$ and $\boldsymbol{B}$ are as above, then:

$$\boldsymbol{AB} = \begin{bmatrix} 6 & 3 \\ -1 & 2 \end{bmatrix}\begin{bmatrix} 1 & 4 \\ 5 & -2 \end{bmatrix}$$

$$= \begin{bmatrix} (6_{11}\times 1_{11}) + (3_{12}\times 5_{21}) & (6_{11}\times 4_{12}) + (3_{12}\times -2_{22}) \\ (-1_{21}\times 1_{11}) + (2_{22}\times 5_{21}) & (-1_{21}\times 4_{12}) + (2_{22}\times -2_{22}) \end{bmatrix} = \begin{bmatrix} 21 & 18 \\ 9 & -8 \end{bmatrix}$$

Subscripts are on the elements in the middle step to indicate which elements should be multiplied and which added. Consider the expression in the first row and column, $(6_{11}\times 1_{11}) + (3_{12}\times 5_{21})$. Notice that for each product, the outside subscripts (i.e., the row designator for the first number and the column designator for the second) indicate the position in the product matrix of the element you are calculating. In this case, 11, the element in the first row and first column. Notice also that the inside numbers for each term of the sum progress from one to two, the number of columns of the first matrix and the number of rows in the second matrix. (It is not much of a progression in this example, but it would be if the matrices were larger.) If we want to calculate the entry for the second row and first column, then we must consider the expression $(-1_{21} \times 1_{11}) + (2_{22} \times 5_{21})$. Again, the outside subscripts for each product indicate the row and column for the entry we are calculating, and the inside subscripts progress from one to two. You can see for yourself that the same pattern holds for the other two entries in the matrix.

There are three other important points to note from this example. First, to multiply two matrices together the number of columns of the first *must* equal the number of rows of the second. If the dimensions match, the matrices are said to be *conformable* and the multiplication is possible. If the dimensions do not match, the matrices are nonconformable and the multiplication is not possible. Second, the resulting matrix, here ***AB***, has the same number of rows as the first matrix and the same number of columns as the second. Third, matrix multiplication is, in general, *not* commutative. Not only is ***AB*** not necessarily equal to ***BA***, but ***BA*** may be of a different order or may not even exist. (Verify for yourself that

$$\boldsymbol{BA} = \begin{bmatrix} 2 & 11 \\ 32 & 11 \end{bmatrix} \neq \boldsymbol{AB}.)$$

Example: If $\boldsymbol{A} = \begin{bmatrix} 1 & 0 \\ -2 & 4 \\ 3 & 1 \end{bmatrix}$ and $\boldsymbol{B} = \begin{bmatrix} 4 & -1 & 1 \\ 2 & 0 & 3 \end{bmatrix}$, then

$$\boldsymbol{AB} = \begin{bmatrix} (1\times 4)+(0\times 2) & (1\times -1)+(0\times 0) & (1\times 1)+(0\times 3) \\ (-2\times 4)+(4\times 2) & (-2\times -1)+(4\times 0) & (-2\times 1)+(4\times 3) \\ (3\times 4)+(1\times 2) & (3\times -1)+(1\times 0) & (3\times 1)+(1\times 3) \end{bmatrix}$$

$$= \begin{bmatrix} 4 & -1 & 1 \\ 0 & 2 & 10 \\ 14 & -3 & 6 \end{bmatrix} \qquad \text{and}$$

$$\boldsymbol{BA} = \begin{bmatrix} (4\times 1)+(-1\times -2)+(1\times 3) & (4\times 0)+(-1\times 4)+(1\times 1) \\ (2\times 1)+(0\times -2)+(3\times 3) & (2\times 0)+(0\times 4)+(3\times 1) \end{bmatrix}$$

$$= \begin{bmatrix} 9 & -3 \\ 11 & 3 \end{bmatrix}$$

6. Matrix multiplication is associative: $\boldsymbol{A}(\boldsymbol{BC}) = (\boldsymbol{AB})\boldsymbol{C}$.
7. Matrix multiplication is distributive over matrix addition: If $\boldsymbol{AB} + \boldsymbol{AC}$ and $\boldsymbol{A}(\boldsymbol{B} + \boldsymbol{C})$ both exist, then $\boldsymbol{AB} + \boldsymbol{AC} = \boldsymbol{A}(\boldsymbol{B} + \boldsymbol{C})$.
8. A matrix is *symmetric* iff $a_{ij} = a_{ji}$ for all pairs (i, j). Both diagonal and identity matrices (below) are forms of symmetric matrices.
9. A *diagonal* matrix is a square matrix of the form

$$\boldsymbol{D} = \begin{bmatrix} d_{11} & 0 & 0 & \ldots & 0 \\ 0 & d_{22} & 0 & \ldots & 0 \\ 0 & 0 & d_{33} & \ldots & 0 \\ \cdot & \cdot & \cdot & \cdot & \cdot \\ \cdot & \cdot & \cdot & \cdot & \cdot \\ \cdot & \cdot & \cdot & \cdot & \cdot \\ 0 & 0 & 0 & \ldots & d_{nn} \end{bmatrix}$$

10. A special form of the diagonal matrix is the *identity* matrix: e.g., $\boldsymbol{I}_3 = \begin{bmatrix} 1 & 0 & 0 \\ 0 & 1 & 0 \\ 0 & 0 & 1 \end{bmatrix}$. The identity matrix is often designated with only one subscript because the number of rows and columns must be equal.

11. The *transpose* of a matrix $\boldsymbol{A}$ is another matrix designated by $\boldsymbol{A}^{\mathrm{T}}$ or $\boldsymbol{A}'$: e.g., if $\boldsymbol{A}_{m\times n} = [a_{ij}]$, then $\boldsymbol{A}^{\mathrm{T}}_{n\times m} = [a_{ji}]$.

Example: If $\boldsymbol{A} = \begin{bmatrix} 0 & -1 & 8 \\ -2 & 3 & 6 \end{bmatrix}$, then $\boldsymbol{A}^{\mathrm{T}} = \begin{bmatrix} 0 & -2 \\ -1 & 3 \\ 8 & 6 \end{bmatrix}$.

6.1.2 Representing a System of Equations in Matrix Form

We can represent a system of *m* linear equations with *n* variables as a matrix equation. Consider the following system of equations:

$$\begin{aligned} a_{11}x_1 + a_{12}x_2 + a_{13}x_3 &= b_1 \\ a_{21}x_1 + a_{22}x_2 + a_{23}x_3 &= b_2 \\ a_{31}x_1 + a_{32}x_2 + a_{33}x_3 &= b_3 \end{aligned}$$

We can think of this system of equations in matrix form as $\boldsymbol{AX} = \boldsymbol{B}$ where

$$\boldsymbol{A} = \begin{bmatrix} a_{11} & a_{12} & a_{13} \\ a_{21} & a_{22} & a_{23} \\ a_{31} & a_{32} & a_{33} \end{bmatrix}, \quad \boldsymbol{X} = \begin{bmatrix} x_1 \\ x_2 \\ x_3 \end{bmatrix}, \text{ and } \boldsymbol{B} = \begin{bmatrix} b_1 \\ b_2 \\ b_3 \end{bmatrix}$$

Matrix form is a useful way of representing and solving large systems of equations, particularly when social scientists use multiple regression or are estimating a system of simultaneous equations.

6.2 Determinants

The *determinant* of a square matrix is a unique scalar associated with that matrix. Determinants are used to calculate inverses of matrices (next section), both of which are used to solve systems of equations such as those found in regression analysis. The determinant of a matrix $\boldsymbol{A}$ is denoted as $|\boldsymbol{A}|$ or $\det \boldsymbol{A}$ (or sometimes $\boldsymbol{D}$ or $\boldsymbol{D}_A$). Corresponding to the previous notation for matrices, for

$$\boldsymbol{A} = \begin{bmatrix} a_{11} & a_{12} \\ a_{21} & a_{22} \end{bmatrix}, \quad |\boldsymbol{A}| = \begin{vmatrix} a_{11} & a_{12} \\ a_{21} & a_{22} \end{vmatrix}$$

The only change in notation is the substitution of vertical lines for the brackets.

For $\boldsymbol{A} = [a_{11}]$, a scalar, $|\boldsymbol{A}| = a_{11}$. For

$$\boldsymbol{A} = \begin{bmatrix} a_{11} & a_{12} \\ a_{21} & a_{22} \end{bmatrix}, \quad |\boldsymbol{A}| = a_{11}a_{22} - a_{12}a_{21}$$

which is the product of the elements on the main diagonal minus the product of the elements on the off-diagonal.

Example: If $\mathbf{A} = \begin{bmatrix} 1 & 2 \\ 3 & 4 \end{bmatrix}$, then $|\mathbf{A}| = \begin{vmatrix} 1 & 2 \\ 3 & 4 \end{vmatrix} = (1 \times 4) - (2 \times 3) = 4 - 6 = -2$.

Example: If $\mathbf{B} = \begin{bmatrix} 2 & 7 \\ -5 & 1 \end{bmatrix}$, then $|\mathbf{B}| = \begin{vmatrix} 2 & 7 \\ -5 & 1 \end{vmatrix} = (2 \times 1) - (7 \times -5) = 2 + 35 = 37$.

The determinant of a third order matrix (i.e., a 3×3 matrix) is a bit more complicated. Suppose

$$\mathbf{A} = \begin{bmatrix} a_{11} & a_{12} & a_{13} \\ a_{21} & a_{22} & a_{23} \\ a_{31} & a_{32} & a_{33} \end{bmatrix}$$

then

$$|\mathbf{A}| = \begin{vmatrix} a_{11} & a_{12} & a_{13} \\ a_{21} & a_{22} & a_{23} \\ a_{31} & a_{32} & a_{33} \end{vmatrix} =$$

$$(a_{11}a_{22}a_{33}) + (a_{12}a_{23}a_{31}) + (a_{13}a_{32}a_{21}) - (a_{31}a_{22}a_{13}) - (a_{32}a_{23}a_{11}) - (a_{33}a_{12}a_{21})$$

There are two common ways of remembering which elements go into each of the six terms of the sum. First, in the "butterfly method" we draw lines connecting the elements of each term. Begin by drawing a line down the main diagonal to connect the elements of $(a_{11}a_{22}a_{33})$. Next draw a line connecting a_{12} and a_{23}. To connect the final element of this term, a_{31}, loop the line around the lower right corner of the array. For the third term, begin the line with a_{13} then loop it around the lower right corner of the array to connect a_{32} and a_{21}. The next three terms are subtracted so we draw the lines in the opposite direction. Begin with a_{31} and draw a line through the diagonal to a_{13}. For the next line, connect a_{32} and a_{23} then loop the line around the upper right corner of the array to connect a_{11}. For the final line, begin with a_{33} then loop the line around the upper right corner of the array to connect it to a_{12} and a_{21}. After drawing all six lines the result should look a bit like a butterfly on its side (or sideways hearts).

The second method is a bit more straightforward. Begin by copying the first two columns of the determinant along the right vertical line as follows:

$$|\boldsymbol{A}| = \left| \begin{array}{ccc|cc} a_{11} & a_{12} & a_{13} & a_{11} & a_{12} \\ a_{21} & a_{22} & a_{23} & a_{21} & a_{22} \\ a_{31} & a_{32} & a_{33} & a_{31} & a_{32} \end{array} \right.$$

Now draw six lines as before connecting the same elements, but without having to loop the lines around the corners of the array.

Determinants of order higher than three require a more formal approach. This more formal method can also be used on determinants of order three (and two), so let's begin with the matrix $\boldsymbol{A}$ from above. We know

$$\begin{aligned} |\boldsymbol{A}| &= (a_{11}a_{22}a_{33}) + (a_{12}a_{23}a_{31}) + (a_{13}a_{32}a_{21}) \\ &\quad - (a_{31}a_{22}a_{13}) - (a_{32}a_{23}a_{11}) - (a_{33}a_{12}a_{21}) \end{aligned}$$

Now reorganize the terms as follows:

$$\begin{aligned} |\boldsymbol{A}| &= (a_{11}a_{22}a_{33}) - (a_{32}a_{23}a_{11}) + (a_{12}a_{23}a_{31}) \\ &\quad - (a_{33}a_{12}a_{21}) + (a_{13}a_{32}a_{21}) - (a_{31}a_{22}a_{13}) \\ &= a_{11}(a_{22}a_{33} - a_{23}a_{32}) + a_{12}(a_{23}a_{31} - a_{21}a_{33}) + a_{13}(a_{21}a_{32} - a_{22}a_{31}) \\ &= a_{11}\begin{vmatrix} a_{22} & a_{23} \\ a_{32} & a_{33} \end{vmatrix} - a_{12}\begin{vmatrix} a_{21} & a_{23} \\ a_{31} & a_{33} \end{vmatrix} + a_{13}\begin{vmatrix} a_{21} & a_{22} \\ a_{31} & a_{32} \end{vmatrix} \end{aligned}$$

In the first line the six terms were rearranged so that the two terms containing a_{11} were first, followed by the two terms containing a_{12}, and finally the two terms containing a_{13}. In the second line, a_{11}, a_{12}, and a_{13} were factored from each respective pair of terms. The value in each set of parentheses looks like the result from taking the determinant of a 2×2 matrix. In the third line, the values in parentheses were put into determinant form. For the second term, -1 was factored from the value. The third line is known as the *expansion* of $|\boldsymbol{A}|$ by its first row. Consider

$$|A| = \begin{vmatrix} a_{11} & a_{12} & a_{13} \\ a_{21} & a_{22} & a_{23} \\ a_{31} & a_{32} & a_{33} \end{vmatrix}$$

again. To expand by the first row we begin with a_{11}. By blocking out all the other elements of the first row and column we are left with

$$\begin{vmatrix} a_{22} & a_{23} \\ a_{32} & a_{33} \end{vmatrix}$$

which is the 2×2 determinant we see above. Next consider a_{12}. This time block out the other elements of the first row and second column and we are left with

$$\begin{vmatrix} a_{21} & a_{23} \\ a_{31} & a_{33} \end{vmatrix}$$

For the final element of the first row, a_{13}, block out the remaining elements of the first row and third column and we are left with

$$\begin{vmatrix} a_{21} & a_{22} \\ a_{31} & a_{32} \end{vmatrix}$$

In other words, in expanding $|A|$ by its first row we obtain three terms (because $|A|$ is of order three), where each term consists of the *j*th element of the first row multiplied by the 2×2 determinant remaining after blocking out the remaining elements of the first row and *j*th column. The sign of each term is $(-1)^{i+j}$. For the first term $(-1)^{1+1} = 1$, so the sign is positive. The sign of the second term is negative because $(-1)^{1+2} = -1$. The sign of the third term will again be positive. In general, the sign of the terms will alternate as one moves across a row or down a column.

For matrix $\mathbf{A}_{n \times n}$, the $(n-1) \times (n-1)$ submatrix formed by deleting the *i*th row and *j*th column is called the *minor* of a_{ij} from $\mathbf{A}$. This minor is denoted by $\mathbf{A}_{ij}$, $\mathbf{M}_{ij}(\mathbf{A})$, or $((a_{ij}))$. The determinant of the a_{ij} minor is denoted by $|\mathbf{A}_{ij}|$, det $\mathbf{A}_{ij}$, det $\mathbf{M}_{ij}(\mathbf{A})$, or det $((a_{ij}))$. We can now state the general formulas for row and column expansions. To expand by any one row, $i = 1, 2, \ldots, n$, use $|\mathbf{A}| = \sum_{j=1}^{n} (-1)^{i+j} a_{ij} |\mathbf{A}_{ij}|$. To expand by any one column, $j = 1, 2, \ldots, n$, use $|\mathbf{A}| = \sum_{i=1}^{n} (-1)^{i+j} a_{ij} |\mathbf{A}_{ij}|$.

Example: Expand the matrix below by the first column:

$$|A| = \begin{vmatrix} a_{11} & a_{12} & a_{13} & a_{14} \\ a_{21} & a_{22} & a_{23} & a_{24} \\ a_{31} & a_{32} & a_{33} & a_{34} \\ a_{41} & a_{42} & a_{43} & a_{44} \end{vmatrix} =$$

$$(-1)^{1+1}a_{11}|A_{11}| + (-1)^{2+1}a_{21}|A_{21}| + (-1)^{3+1}a_{31}|A_{31}| + (-1)^{4+1}a_{41}|A_{41}|$$

where

$$|A_{11}| = \begin{vmatrix} a_{22} & a_{23} & a_{24} \\ a_{32} & a_{33} & a_{34} \\ a_{42} & a_{43} & a_{44} \end{vmatrix}, \quad |A_{21}| = \begin{vmatrix} a_{12} & a_{13} & a_{14} \\ a_{32} & a_{33} & a_{34} \\ a_{42} & a_{43} & a_{44} \end{vmatrix}$$

$$|A_{31}| = \begin{vmatrix} a_{12} & a_{13} & a_{14} \\ a_{22} & a_{23} & a_{24} \\ a_{42} & a_{43} & a_{44} \end{vmatrix}, \quad |A_{41}| = \begin{vmatrix} a_{12} & a_{13} & a_{14} \\ a_{22} & a_{23} & a_{24} \\ a_{32} & a_{33} & a_{34} \end{vmatrix}$$

I will leave it to you to complete the expansion by calculating each of the $|A_{i1}|$.

As you can see, even with just a 4×4 determinant, row or column expansion can become rather messy. You can make the process easier by selecting the row or column with the most zeros and ones. Fortunately, we can also manipulate the rows and columns to create zeros and ones. In particular, the value of $|A|$ does not change if we replace one row (or column) by itself plus k times another row (or column). (*Note:* you cannot mix rows and columns in the same step.)

Example: Find $|A| = \begin{vmatrix} 1 & 2 & 3 \\ 2 & 4 & 1 \\ 1 & 3 & 0 \end{vmatrix}$. It would be relatively easy to expand by the first or third columns, but let's begin by subtracting twice the first column from the second:

$$|A| = \begin{vmatrix} 1 & 2-2 & 3 \\ 2 & 4-4 & 1 \\ 1 & 3-2 & 0 \end{vmatrix} = \begin{vmatrix} 1 & 0 & 3 \\ 2 & 0 & 1 \\ 1 & 1 & 0 \end{vmatrix}$$

Now we can easily expand by the second column as follows:

$$|\boldsymbol{A}| = (-1)^{1+2}0\begin{vmatrix} 2 & 1 \\ 1 & 0 \end{vmatrix} + (-1)^{2+2}0\begin{vmatrix} 1 & 3 \\ 1 & 0 \end{vmatrix} + (-1)^{3+2}1\begin{vmatrix} 1 & 3 \\ 2 & 1 \end{vmatrix}$$

$$= 0 + 0 + (-1)^{3+2}1\begin{vmatrix} 1 & 3 \\ 2 & 1 \end{vmatrix}$$

$$= (-1)[(1 \times 1) - (2 \times 3)] = (-1)(-5) = 5$$

Because the first two elements of the second column were zeros, we could have left out the first two terms and just started with calculations on the third line.

Example: Evaluate $|\boldsymbol{B}| = \begin{vmatrix} 2 & 1 & -3 & 4 \\ 5 & -4 & 7 & -2 \\ 4 & 0 & 6 & -3 \\ 3 & -2 & 5 & -2 \end{vmatrix}$.

Because the second column has a one and a zero, let us work with it. Start by adding four times the first row to the second:

$$|\boldsymbol{B}| = \begin{vmatrix} 2 & 1 & -3 & 4 \\ 5+8 & -4+4 & 7+(-12) & -2+16 \\ 4 & 0 & 6 & -3 \\ 3 & -2 & 5 & -2 \end{vmatrix} = \begin{vmatrix} 2 & 1 & -3 & 4 \\ 13 & 0 & -5 & 14 \\ 4 & 0 & 6 & -3 \\ 3 & -2 & 5 & -2 \end{vmatrix}$$

Now add two times the first row to the fourth:

$$|\boldsymbol{B}| = \begin{vmatrix} 2 & 1 & -3 & 4 \\ 13 & 0 & -5 & 14 \\ 4 & 0 & 6 & -3 \\ 3+4 & -2+2 & 5+(-6) & -2+8 \end{vmatrix} = \begin{vmatrix} 2 & 1 & -3 & 4 \\ 13 & 0 & -5 & 14 \\ 4 & 0 & 6 & -3 \\ 7 & 0 & -1 & 6 \end{vmatrix}$$

Now expand by the second column:

$$|\boldsymbol{B}| = (-1)^{1+2}(1)\begin{vmatrix} 13 & -5 & 14 \\ 4 & 6 & -3 \\ 7 & -1 & 6 \end{vmatrix}$$

We could now apply the butterfly method, but instead continue manipulating the rows so we can expand by the second column of the remaining determinant. Begin by adding six times the third row to the second:

$$|\boldsymbol{B}| = (-1)\begin{vmatrix} 13 & -5 & 14 \\ 4+42 & 6+(-6) & -3+36 \\ 7 & -1 & 6 \end{vmatrix} = (-1)\begin{vmatrix} 13 & -5 & 14 \\ 46 & 0 & 33 \\ 7 & -1 & 6 \end{vmatrix}$$

Next subtract five times the third row from the first:

$$|\boldsymbol{B}| = (-1)\begin{vmatrix} 13-35 & -5-(-5) & 14-30 \\ 46 & 0 & 33 \\ 7 & -1 & 6 \end{vmatrix} = (-1)\begin{vmatrix} -22 & 0 & -16 \\ 46 & 0 & 33 \\ 7 & -1 & 6 \end{vmatrix}$$

Now expand by the second column:

$$|\boldsymbol{B}| = (-1)\left((-1)^{3+2}(-1)\begin{vmatrix} -22 & -16 \\ 46 & 33 \end{vmatrix}\right)$$

The first (–1) is from the first expansion and everything inside the large parentheses is from the second. Finishing the calculations we find:

$$|\boldsymbol{B}| = (-1)\begin{vmatrix} -22 & -16 \\ 46 & 33 \end{vmatrix} =$$

$$(-1)[(-22 \times 33) - (-16 \times 46)] = (-1)[-726 - (-736)] = (-1)(10) = -10$$

Some final comments on determinants:

1. Matrices and determinants are not the same thing. A determinant is a scalar defined for a square matrix.
2. The determinant of $\boldsymbol{A}$ is equal to the determinant of the transpose of $\boldsymbol{A}$: $|\boldsymbol{A}| = |\boldsymbol{A}^{\mathrm{T}}|$. In addition, if any two rows of a matrix are interchanged then the sign of the determinant changes.
3. The determinant of the product of two square matrices is equal to the product of their determinants: $|\boldsymbol{AB}| = |\boldsymbol{A}| \times |\boldsymbol{B}|$. (Of course $\boldsymbol{AB}$ could be square without either $\boldsymbol{A}$ or $\boldsymbol{B}$ being square, but if that were true, neither $|\boldsymbol{A}|$ nor $|\boldsymbol{B}|$ would be defined.)

4. If a single row or column of a matrix is multiplied by a constant, the value of the determinant of the new matrix is equal to the determinant of the original matrix times the constant: If $A = \begin{bmatrix} a & b \\ c & d \end{bmatrix}$, and $B = \begin{bmatrix} ka & kb \\ c & d \end{bmatrix}$, then $|B| = (ka \times d) - (kb \times c) = kad - kbc = k(ad - bc) = k|A|$.
5. If a matrix is multiplied by a constant, then the determinant of the new matrix is equal to the determinant of the original matrix times the constant raised to the *n*th power, where *n* is the number of rows (or columns because it is square) in the matrix: If $A = \begin{bmatrix} a & b \\ c & d \end{bmatrix}$ and $C = kA = \begin{bmatrix} ka & kb \\ kc & kd \end{bmatrix}$, then $|C| = \begin{vmatrix} ka & kb \\ kc & kd \end{vmatrix} = (ka \times kd) - (kb \times kc) = k^2ad - k^2bc = k^2(ad - bc) = k^2|A|$.

6.3 Inverses of Matrices

The *inverse* of an $n \times n$ matrix A, if it exists, is denoted A^{-1} and is also an $n \times n$ matrix such that $A^{-1}A = AA^{-1} = I_n$. (In a sense, multiplication by an inverse is the closest we can come to division in matrix operations.) Inverses of matrices are important to social scientists because they can be used to solve systems of equations. Recall the system of equations presented in section 6.1.2. If $AX = B$, where A is a matrix of values for the independent variables, B is a vector of values for the dependent variable, and X is a vector of coefficients for the variables, then we can solve for X by premultiplying both sides of the equation by A^{-1}:

$$A^{-1}AX = A^{-1}B \Rightarrow IX = A^{-1}B \Rightarrow X = A^{-1}B$$

To determine the inverse of a matrix A, begin by calculating the determinant $|A|$. If $|A| \neq 0$, then the matrix is said to be *nonsingular*. If $|A| = 0$, the matrix is singular, or is said to have a singularity, which means one or more rows or columns is a linear transformation of another row or column. In other words, the rows and columns are not independent. Such a lack of independence may indicate a modeling error in the system of equations, but it also means an inverse for that matrix cannot be calculated and, thus, that there is no unique solution to the system of equations. (*Note:*

If A is singular, we may still want to know how much of A is linearly independent. The *rank* of a matrix is the order of the largest square submatrix of A whose determinant is not equal to zero. The rank of a matrix is important when working with simultaneous equations.)

If the matrix A is nonsingular, we next determine the *adjoint* of A, denoted adj A. The adjoint of A is the transpose of the cofactor matrix of A: adj $A = (\text{cof}\,A)^T$. Recall that $|A| = \sum_{j=1}^{n} (-1)^{i+j} a_{ij} |A_{ij}|$, for any one row, is the row expansion of $|A|$. The *cofactor* or *signed minor* of the element a_{ij} is a scalar given by $|C_{ij}| = (-1)^{i+j} |A_{ij}|$. (*Note:* You do not multiply by a_{ij} to find the cofactor.) The *cofactor matrix* of A is the matrix formed by the cofactors of all the elements of A and is denoted cof A, $[|C_{ij}|]$, or just C. Thus if

$$A = \begin{bmatrix} a_{11} & a_{12} & \cdots & a_{1n} \\ a_{21} & a_{22} & \cdots & a_{2n} \\ \cdot & \cdot & & \cdot \\ \cdot & \cdot & & \cdot \\ \cdot & \cdot & & \cdot \\ a_{n1} & a_{n2} & \cdots & a_{nn} \end{bmatrix}$$

then

$$\text{cof}\,A = \begin{bmatrix} |C_{11}| & |C_{12}| & \cdots & |C_{1n}| \\ |C_{21}| & |C_{22}| & \cdots & |C_{2n}| \\ \cdot & \cdot & & \cdot \\ \cdot & \cdot & & \cdot \\ \cdot & \cdot & & \cdot \\ |C_{n1}| & |C_{n2}| & \cdots & |C_{nn}| \end{bmatrix}$$

and

$$\text{adj}\,A = \begin{bmatrix} |C_{11}| & |C_{21}| & \cdots & |C_{n1}| \\ |C_{12}| & |C_{22}| & \cdots & |C_{n2}| \\ \cdot & \cdot & & \cdot \\ \cdot & \cdot & & \cdot \\ \cdot & \cdot & & \cdot \\ |C_{1n}| & |C_{2n}| & \cdots & |C_{nn}| \end{bmatrix}$$

Once we know the adjoint of a square nonsingular matrix $\boldsymbol{A}$, we can calculate its inverse as

$$\boldsymbol{A}^{-1} = \left(\frac{1}{|\boldsymbol{A}|}\right)\text{adj}\boldsymbol{A}.$$

(It should be clear from this formula why we cannot calculate an inverse for a singular matrix, i.e., when $|\boldsymbol{A}| = 0$.)

Example: Find the inverse of $\boldsymbol{A} = \begin{bmatrix} 1 & 2 \\ 3 & 4 \end{bmatrix}$. First note that the matrix is square and $|\boldsymbol{A}| = -2$. Next determine that $\text{cof}\boldsymbol{A} = \begin{bmatrix} 4 & -3 \\ -2 & 1 \end{bmatrix}$ and adj $\boldsymbol{A}$ = $(\text{cof}\boldsymbol{A})^{\text{T}} = \begin{bmatrix} 4 & -2 \\ -3 & 1 \end{bmatrix}$. Then:

$$\boldsymbol{A}^{-1} = \left(\frac{1}{-2}\right)\begin{bmatrix} 4 & -2 \\ -3 & 1 \end{bmatrix} = \begin{bmatrix} 4/{-2} & -2/{-2} \\ -3/{-2} & 1/{-2} \end{bmatrix} = \begin{bmatrix} -2 & 1 \\ 3/2 & -1/2 \end{bmatrix}.$$

If this answer is correct, $\boldsymbol{A}^{-1}\boldsymbol{A} = \boldsymbol{A}\boldsymbol{A}^{-1} = \boldsymbol{I}_2$:

$$\boldsymbol{A}^{-1}\boldsymbol{A} = \begin{bmatrix} -2 & 1 \\ 3/2 & -1/2 \end{bmatrix}\begin{bmatrix} 1 & 2 \\ 3 & 4 \end{bmatrix} = \begin{bmatrix} (-2 \times 1) + (1 \times 3) & (-2 \times 2) + (1 \times 4) \\ (3/2 \times 1) + (-1/2 \times 3) & (3/2 \times 2) + (-1/2 \times 4) \end{bmatrix}$$

$$= \begin{bmatrix} -2+3 & -4+4 \\ 3/2 - 3/2 & 3-2 \end{bmatrix} = \begin{bmatrix} 1 & 0 \\ 0 & 1 \end{bmatrix}$$

$$\boldsymbol{A}\boldsymbol{A}^{-1} = \begin{bmatrix} 1 & 2 \\ 3 & 4 \end{bmatrix}\begin{bmatrix} -2 & 1 \\ 3/2 & -1/2 \end{bmatrix} = \begin{bmatrix} (1 \times -2) + (2 \times 3/2) & (1 \times 1) + (2 \times -1/2) \\ (3 \times -2) + (4 \times 3/2) & (3 \times 1) + (4 \times -1/2) \end{bmatrix}$$

$$= \begin{bmatrix} -2+3 & 1-1 \\ -6+6 & 3-2 \end{bmatrix} = \begin{bmatrix} 1 & 0 \\ 0 & 1 \end{bmatrix}$$

Example: Find the inverse of $\boldsymbol{B} = \begin{bmatrix} 1 & 2 & 3 \\ 2 & 4 & 5 \\ 3 & 5 & 6 \end{bmatrix}$ if it exists. Begin by determining that $|\boldsymbol{B}| = -1$. Next determine the cofactor matrix of $\boldsymbol{B}$:

$$\text{cof}\,\boldsymbol{B} =$$

$$\begin{bmatrix} (-1)^{1+1}[(4\times 6)-(5\times 5)] & (-1)^{1+2}[(2\times 6)-(5\times 3)] & (-1)^{1+3}[(2\times 5)-(4\times 3)] \\ (-1)^{2+1}[(2\times 6)-(3\times 5)] & (-1)^{2+2}[(1\times 6)-(3\times 3)] & (-1)^{2+3}[(1\times 5)-(2\times 3)] \\ (-1)^{3+1}[(2\times 5)-(3\times 4)] & (-1)^{3+2}[(1\times 5)-(3\times 2)] & (-1)^{3+3}[(1\times 4)-(2\times 2)] \end{bmatrix}$$

$$= \begin{bmatrix} -1 & 3 & -2 \\ 3 & -3 & 1 \\ -2 & 1 & 0 \end{bmatrix}$$

Then

$$\text{adj}\,\boldsymbol{B} = \begin{bmatrix} -1 & 3 & -2 \\ 3 & -3 & 1 \\ -2 & 1 & 0 \end{bmatrix} \quad \text{and} \quad \boldsymbol{B}^{-1} = (-1)\begin{bmatrix} -1 & 3 & -2 \\ 3 & -3 & 1 \\ -2 & 1 & 0 \end{bmatrix} = \begin{bmatrix} 1 & -3 & 2 \\ -3 & 3 & -1 \\ 2 & -1 & 0 \end{bmatrix}.$$

I will leave it for you to check that $\boldsymbol{B}^{-1}\boldsymbol{B} = \boldsymbol{B}\boldsymbol{B}^{-1} = \boldsymbol{I}_3$.

More Matrix Rules: Now that we have defined the inverse of a matrix we can list some additional rules for matrices:

1. $k(\boldsymbol{A}+\boldsymbol{B}) = k\boldsymbol{A} + k\boldsymbol{B}$
2. $(\boldsymbol{A}^{\mathrm{T}})^{\mathrm{T}} = \boldsymbol{A}$
3. $(\boldsymbol{A}+\boldsymbol{B})^{\mathrm{T}} = \boldsymbol{A}^{\mathrm{T}} + \boldsymbol{B}^{\mathrm{T}}$
4. $(\boldsymbol{AB})^{\mathrm{T}} = \boldsymbol{B}^{\mathrm{T}}\boldsymbol{A}^{\mathrm{T}}$
5. $(\boldsymbol{ABC})^{\mathrm{T}} = \boldsymbol{C}^{\mathrm{T}}\boldsymbol{B}^{\mathrm{T}}\boldsymbol{A}^{\mathrm{T}}$
6. $(\boldsymbol{AB})^{-1} = \boldsymbol{B}^{-1}\boldsymbol{A}^{-1}$
7. $(\boldsymbol{ABC})^{-1} = \boldsymbol{C}^{-1}\boldsymbol{B}^{-1}\boldsymbol{A}^{-1}$
8. $(\boldsymbol{A}^{-1})^{-1} = \boldsymbol{A}$
9. $(\boldsymbol{A}^{\mathrm{T}})^{-1} = (\boldsymbol{A}^{-1})^{\mathrm{T}}$

6.4 Cramer's Rule

We have seen how we can represent a system of linear equations in matrix form, $\boldsymbol{A}_{n\times n}\boldsymbol{X}_{n\times 1} = \boldsymbol{B}_{n\times 1}$, and solve for

$$X_{n\times 1} = A^{-1}B = \left(\frac{1}{|A|}\right)(\text{adj}\,A)B.$$

In a more elaborate form, the solution for X is:

$$\begin{bmatrix} x_1 \\ x_2 \\ \cdot \\ \cdot \\ \cdot \\ x_n \end{bmatrix} = \left(\frac{1}{|A|}\right) \begin{bmatrix} |C_{11}| & |C_{21}| & \cdots & |C_{n1}| \\ |C_{12}| & |C_{22}| & \cdots & |C_{n2}| \\ \cdot & \cdot & & \cdot \\ \cdot & \cdot & & \cdot \\ \cdot & \cdot & & \cdot \\ |C_{1n}| & |C_{2n}| & \cdots & |C_{nn}| \end{bmatrix} \begin{bmatrix} b_1 \\ b_2 \\ \cdot \\ \cdot \\ \cdot \\ b_n \end{bmatrix}$$

$$= \left(\frac{1}{|A|}\right) \begin{bmatrix} b_1|C_{11}| + b_2|C_{21}| + \ldots + b_n|C_{n1}| \\ b_1|C_{12}| + b_2|C_{22}| + \ldots + b_n|C_{n2}| \\ \cdot \\ \cdot \\ \cdot \\ b_1|C_{1n}| + b_2|C_{2n}| + \ldots + b_n|C_{nn}| \end{bmatrix}$$

Notice that the dimensions of the matrix in the last term are $n \times 1$. This means that

$$x_j = \left(\frac{1}{|A|}\right) \sum_{i=1}^{n} b_i\,|C_{ij}|$$

The quantity $\sum_{i=1}^{n} b_i\,|C_{ij}|$ is very much like taking a determinant. Cramer's rule tells us that to solve for x_j in a set of simultaneous equations we replace the jth column of A with B, take the determinant of this new matrix, and divide it by the determinant of A; that is,

$$x_j = \left(\frac{1}{|A|}\right) |A_{(j\leftrightarrow B)}|,$$

where $A_{(j\leftrightarrow B)}$ is A with its jth column replaced by B. (*Note:* Because $X_{n\times 1}$ is a column vector its elements would normally be represented as

x_i, but I use x_j here to remind you that the jth column of $\boldsymbol{A}$ will be replaced by $\boldsymbol{B}$.)

Example: Use Cramer's rule to solve this system of equations: $x_1 + 2x_2 - x_3 = 3$, $-x_1 + x_2 + 2x_3 = 2$, and $2x_1 - x_2 + x_3 = 1$. Begin by putting the equations into matrix form.

$$\begin{aligned} x_1 + 2x_2 - x_3 &= 3 \\ -x_1 + x_2 + 2x_3 &= 2 \\ 2x_1 - x_2 + x_3 &= 1 \end{aligned} \Rightarrow \boldsymbol{AX} = \boldsymbol{B} \Rightarrow \begin{bmatrix} 1 & 2 & -1 \\ -1 & 1 & 2 \\ 2 & -1 & 1 \end{bmatrix} \begin{bmatrix} x_1 \\ x_2 \\ x_3 \end{bmatrix} = \begin{bmatrix} 3 \\ 2 \\ 1 \end{bmatrix}$$

To solve for x_1 we substitute $\boldsymbol{B}$ for the first column of $\boldsymbol{A}$, take the determinant of the new matrix, and divide by the determinant of $\boldsymbol{A}$:

$$x_1 = \frac{|\boldsymbol{A}_{(1 \leftrightarrow \boldsymbol{B})}|}{|\boldsymbol{A}|} = \frac{\begin{vmatrix} 3 & 2 & -1 \\ 2 & 1 & 2 \\ 1 & -1 & 1 \end{vmatrix}}{\begin{vmatrix} 1 & 2 & -1 \\ -1 & 1 & 2 \\ 2 & -1 & 1 \end{vmatrix}} = \frac{12}{14} = \frac{6}{7}$$

Next calculate x_2 and x_3:

$$x_2 = \frac{|\boldsymbol{A}_{(2 \leftrightarrow \boldsymbol{B})}|}{|\boldsymbol{A}|} = \frac{\begin{vmatrix} 1 & 3 & -1 \\ -1 & 2 & 2 \\ 2 & 1 & 1 \end{vmatrix}}{14} = \frac{20}{14} = \frac{10}{7}$$

$$x_3 = \frac{|\boldsymbol{A}_{(3 \leftrightarrow \boldsymbol{B})}|}{|\boldsymbol{A}|} = \frac{\begin{vmatrix} 1 & 2 & 3 \\ -1 & 1 & 2 \\ 2 & -1 & 1 \end{vmatrix}}{14} = \frac{10}{14} = \frac{5}{7}$$

To verify these answers insert them into the equations:

$$\frac{6}{7}+2\left(\frac{10}{7}\right)-\frac{5}{7}=\frac{6+20-5}{7}=3$$

$$-\frac{6}{7}+\frac{10}{7}+2\left(\frac{5}{7}\right)=\frac{-6+10+10}{7}=2$$

$$2\left(\frac{6}{7}\right)-\frac{10}{7}+\frac{5}{7}=\frac{12-10+5}{7}=1$$

6.5 Eigenvalues and Eigenvectors

$\boldsymbol{X}$ is called an *eigenvector* (characteristic vector; *eigen* is German for *characteristic*) if there exists a nonzero vector $\boldsymbol{X}_{n\times 1}$ such that $\boldsymbol{A}_{n\times n}\boldsymbol{X}_{n\times 1} = \lambda\boldsymbol{X}_{n\times 1}$. The scalar λ is called an *eigenvalue* of $\boldsymbol{A}_{n\times n}$. If λ is an eigenvalue of $\boldsymbol{A}$, then $\boldsymbol{AX}=\lambda\boldsymbol{X}$ for some $\boldsymbol{X}$. Then it follows that $\boldsymbol{AX}-\lambda\boldsymbol{X}=\boldsymbol{0}$ and $(\boldsymbol{A}-\lambda\boldsymbol{I})\boldsymbol{X}=\boldsymbol{0}$, where $\boldsymbol{0}$ is an $n\times n$ matrix whose elements are all 0. We determine the eigenvalues for $\boldsymbol{A}$ by solving $|\boldsymbol{A}-\lambda\boldsymbol{I}| = 0$ for λ. The function $f(\lambda) = |\boldsymbol{A}-\lambda\boldsymbol{I}|$ is called the *characteristic function* of $\boldsymbol{A}$. (When expressed as a polynomial it is called the *characteristic equation* of $\boldsymbol{A}$.) Each eigenvalue has an eigenvector associated with it. To determine the eigenvectors, we substitute each eigenvalue into $(\boldsymbol{A}-\lambda\boldsymbol{I})\boldsymbol{X}=\boldsymbol{0}$ and solve for $\boldsymbol{X}$.

Example: Find the eigenvalues and eigenvectors of $\boldsymbol{A} = \begin{bmatrix} 3 & 1 \\ 2 & 2 \end{bmatrix}$. We begin by constructing $|\boldsymbol{A}-\lambda\boldsymbol{I}|$:

$$\begin{vmatrix} 3 & 1 \\ 2 & 2 \end{vmatrix} - \lambda\begin{vmatrix} 1 & 0 \\ 0 & 1 \end{vmatrix} = \begin{vmatrix} 3 & 1 \\ 2 & 2 \end{vmatrix} - \begin{vmatrix} \lambda & 0 \\ 0 & \lambda \end{vmatrix} = \begin{vmatrix} 3-\lambda & 1 \\ 2 & 2-\lambda \end{vmatrix}.$$

Now solve $|\boldsymbol{A}-\lambda\boldsymbol{I}| = 0$:

$$\begin{vmatrix} 3-\lambda & 1 \\ 2 & 2-\lambda \end{vmatrix} = (3-\lambda)(2-\lambda) - 1\times 2 = 6 - 5\lambda + \lambda^2 - 2 = \lambda^2 - 5\lambda + 4 = 0$$

$$\Rightarrow (\lambda-4)(\lambda-1)=0, \ *\lambda = 1, 4$$

Using $(\boldsymbol{A}-\lambda\boldsymbol{I})\boldsymbol{X}=\boldsymbol{0}$ we can determine the eigenvector associated with each eigenvalue. For $\lambda = 1$:

$$\begin{bmatrix} 3-1 & 1 \\ 2 & 2-1 \end{bmatrix}\begin{bmatrix} x_1 \\ x_2 \end{bmatrix} = \begin{bmatrix} 0 \\ 0 \end{bmatrix} \Rightarrow \begin{bmatrix} 2 & 1 \\ 2 & 1 \end{bmatrix}\begin{bmatrix} x_1 \\ x_2 \end{bmatrix} = \begin{bmatrix} 0 \\ 0 \end{bmatrix} \Rightarrow \begin{bmatrix} 2x_1 + x_2 \\ 2x_1 + x_2 \end{bmatrix} = \begin{bmatrix} 0 \\ 0 \end{bmatrix}$$

We do not obtain specific values for the x_i. We know that $2x_1 + x_2 = 0 \Rightarrow x_1 = -\frac{1}{2}x_2$, so if $x_2 = k$, then $x_1 = -\frac{1}{2}k$. (Here, x_1 is in terms of x_2, but either way is acceptable.) Thus the eigenvector for $\lambda = 1$ is

$$\begin{bmatrix} -\frac{1}{2}k \\ k \end{bmatrix}$$

Check these answers using $(\boldsymbol{A} - 1 \times \boldsymbol{I})\boldsymbol{X} = \boldsymbol{0}$:

$$\begin{bmatrix} 3-1 & 1 \\ 2 & 2-1 \end{bmatrix} \begin{bmatrix} -\frac{1}{2}k \\ k \end{bmatrix} = \begin{bmatrix} 2 & 1 \\ 2 & 1 \end{bmatrix} \begin{bmatrix} -\frac{1}{2}k \\ k \end{bmatrix} = \begin{bmatrix} 2(-\frac{1}{2}k) + k \\ 2(-\frac{1}{2}k) + k \end{bmatrix} = \begin{bmatrix} 0 \\ 0 \end{bmatrix}$$

Now determine the eigenvector for $\lambda = 4$. As before, begin with $(\boldsymbol{A} - 4\boldsymbol{I})\boldsymbol{X} = \boldsymbol{0}$:

$$\begin{bmatrix} 3-4 & 1 \\ 2 & 2-4 \end{bmatrix} \begin{bmatrix} x_1 \\ x_2 \end{bmatrix} = \begin{bmatrix} -1 & 1 \\ 2 & -2 \end{bmatrix} \begin{bmatrix} x_1 \\ x_2 \end{bmatrix} = \begin{bmatrix} -x_1 + x_2 \\ 2x_1 - 2x_2 \end{bmatrix} = \begin{bmatrix} 0 \\ 0 \end{bmatrix}$$

This time the two equations are different, but only by a factor of –2. Using the first equation we find that $x_1 = x_2$, so if $x_2 = k$, then $x_1 = k$, and the eigenvector for $\lambda = 4$ is

$$\begin{bmatrix} k \\ k \end{bmatrix}$$

Check these answers using $(\boldsymbol{A} - 4\boldsymbol{I})\boldsymbol{X} = \boldsymbol{0}$:

$$\begin{bmatrix} 3-4 & 1 \\ 2 & 2-4 \end{bmatrix} \begin{bmatrix} k \\ k \end{bmatrix} = \begin{bmatrix} -1 & 1 \\ 2 & -2 \end{bmatrix} \begin{bmatrix} k \\ k \end{bmatrix} = \begin{bmatrix} -k + k \\ 2k - 2k \end{bmatrix} = \begin{bmatrix} 0 \\ 0 \end{bmatrix}.$$

Some Properties of Eigenvalues and Eigenvectors:

1. The sum of the eigenvalues of a matrix is equal to the trace of the matrix. The *trace* of a matrix is the sum of its diagonal elements. For example, the trace of the matrix $\boldsymbol{A}$ in the previous example is 5 = 3 + 2 and the sum of the eigenvalues of that matrix was 1 + 4 = 5.
2. The product of the eigenvalues of a matrix is equal to the determinant of that matrix. The determinant of the matrix of the previous

example was $(3 \times 2) - (1 \times 2) = 4$ and the product of the eigenvalues was $1 \times 4 = 4$.

3. For a nonsingular matrix, the number of eigenvalues (counting duplicates) is equal to the rank of the matrix. In the above example, the rank of the matrix was two and we found two eigenvalues.

Eigenvalues and eigenvectors are important to social scientists. Many applications of matrix algebra involve mappings involving the matrix $(\boldsymbol{A} - \lambda_i \boldsymbol{I})\boldsymbol{X}$, where λ_i is an eigenvalue. Eigenvalues and eigenvectors are also used in factor analysis and regression diagnostics. For example, a singular matrix will signal a linear dependency in the independent variables, but we usually do not want near dependencies either. The size of the eigenvalues will tell us how closely the variables are related.

6.6 Multivariate Extrema and Matrix Algebra

To conclude this chapter we will briefly examine one way matrix algebra is used to find the extrema of a function. Just as we have seen previously in looking for extrema in multivariate functions, we begin by requiring that all the first order partial derivatives be set equal to zero and solved simultaneously. Thus, if $y = f(x_1, x_2, \ldots, x_n)$, then any extreme point must meet the condition: $f_{x_1} = f_{x_2} = \ldots = f_{x_n} = 0$, where f_{x_i} is the first order partial derivative of f with respect to x_i. This is called the *first order (necessary) condition.*

After satisfying the first order condition, we must satisfy the *second order (sufficient) condition.* To do so we must construct the *Hessian determinant* of the function. The Hessian determinant of a function, $y = f(x_1, x_2, \ldots, x_n)$, is denoted by $|\boldsymbol{H}|$ and comprises the second order partial derivatives of the function as follows:

$$|\boldsymbol{H}| = \begin{vmatrix} f_{11} & f_{12} & \cdots & f_{1n} \\ f_{21} & f_{22} & \cdots & f_{2n} \\ \cdot & \cdot & \cdot & \cdot \\ \cdot & \cdot & \cdot & \cdot \\ \cdot & \cdot & \cdot & \cdot \\ f_{n1} & f_{n2} & \cdots & f_{nn} \end{vmatrix}$$

where $f_{ij} = f_{x_i x_j}$.

The Hessian determinant is symmetric. The main diagonal of the Hessian determinant consists of all second order partial derivatives of the function. The off-diagonal elements consist of the mixed partial derivatives for all the variables. The symmetry of the Hessian determinant is due to Young's Theorem (Chapter 4) which tells us that $f_{ij} = f_{ji}$.

We use $|H_i|$ to denote the ith *principal minor* of $|H|$. For example,

$$|H_1| = |f_{11}| = f_{11}, \quad |H_2| = \begin{vmatrix} f_{11} & f_{12} \\ f_{21} & f_{22} \end{vmatrix} = \begin{vmatrix} f_{11} & f_{12} \\ f_{12} & f_{22} \end{vmatrix} = (f_{11} \times f_{22}) - (f_{12})^2,$$

and so on. Once we have found the Hessian determinant and its principal minors, one of these conditions must hold:

1. When all the principal minors are greater than zero (i.e., $|H_1| > 0$, $|H_2| > 0, \ldots, |H_n| > 0$), the critical point is a minimum.
2. When the sign of the principal minors alternates beginning with a negative sign (i.e., $|H_1| < 0$, $|H_2| > 0$, $|H_3| < 0, \ldots$), the critical point is a maximum.
3. When neither of the first two conditions holds, the test fails and we must examine a neighborhood around the critical point to determine whether an extremum exists there.

Example: Find the extreme points of $f(x_1, x_2, x_3) = x_1^2 + 2x_1 + x_2^2 + x_3^2 - 4x_3 + 2$. Begin with the first order condition:

$$f_{x_1} = 2x_1 + 2 = 0$$
$$f_{x_2} = 2x_2 = 0$$
$$f_{x_3} = 2x_3 - 4 = 0$$

We could use Cramer's rule to solve this system of equations, but for this system it is easier to just notice that the second equation tells us $x_2 = 0$. From the first equation we get $x_1 = -1$, and from the third, $x_3 = 2$. Entering these values into $f(x_1, x_2, x_3)$ tells us that the critical point is $(-1, 0, 2, -3)$. Next find the Hessian determinant and the principal minors:

$$|H| = \begin{vmatrix} f_{11} & f_{12} & f_{13} \\ f_{12} & f_{22} & f_{23} \\ f_{13} & f_{23} & f_{33} \end{vmatrix} = \begin{vmatrix} 2 & 0 & 0 \\ 0 & 2 & 0 \\ 0 & 0 & 2 \end{vmatrix}$$

$$|\boldsymbol{H}_1| = |2| = 2, \quad |\boldsymbol{H}_2| = \begin{vmatrix} 2 & 0 \\ 0 & 2 \end{vmatrix} = 4, \quad |\boldsymbol{H}_3| = \begin{vmatrix} 2 & 0 & 0 \\ 0 & 2 & 0 \\ 0 & 0 & 2 \end{vmatrix} = 8$$

Because $|\boldsymbol{H}_1| > 0$, $|\boldsymbol{H}_2| > 0$, and $|\boldsymbol{H}_3| > 0$, the critical point must be a minimum (as we might have guessed from the function itself).

Just as we saw previously with Lagrange multipliers, we often wish to optimize a function subject to some constraint. To optimize $f(x_1, x_2, \ldots, x_n)$ subject to the constraints $g_1(x_1, x_2, \ldots, x_n) = 0$ to $g_m(x_1, x_2, \ldots, x_n) = 0$, where $m < n$, we begin by constructing the function:

$$F(x_1, x_2, \ldots, x_n, \lambda_1, \lambda_2, \ldots, \lambda_m) =$$
$$f(x_1, x_2, \ldots, x_n) - \lambda_1[g_1(x_1, x_2, \ldots, x_n)] \ldots - \lambda_m[g_m(x_1, x_2, \ldots, x_n)].$$

We then solve the first order condition by differentiating with respect to each of the variables (including the λ_i), setting them all equal to zero, and solving all $n + m$ equations simultaneously. For the second order condition we must construct a *bordered Hessian determinant*, which is denoted $|\overline{\boldsymbol{H}}|$. A bordered Hessian determinant has this general structure:

$$|\overline{\boldsymbol{H}}| = \begin{vmatrix} 0 & \ldots & 0 & g_{1,x_1} & g_{1,x_2} & \cdots & g_{1,x_n} \\ \vdots & \ddots & \vdots & \vdots & \vdots & \ddots & \vdots \\ 0 & \ldots & 0 & g_{m,x_1} & g_{m,x_2} & \cdots & g_{m,x_n} \\ g_{1,x_1} & \cdots & g_{m,x_1} & F_{x_1x_1} & F_{x_2x_1} & \cdots & F_{x_nx_1} \\ g_{1,x_2} & \cdots & g_{m,x_2} & F_{x_2x_1} & F_{x_2x_2} & \cdots & F_{x_nx_2} \\ \vdots & \vdots & \vdots & \vdots & \vdots & \vdots & \vdots \\ g_{1,x_n} & \cdots & g_{m,x_n} & F_{x_nx_1} & F_{x_nx_2} & \cdots & F_{x_nx_n} \end{vmatrix}$$

There are four basic parts to the structure. The upper left section consists of an $m \times m$ array of zeros. The upper right and lower left sections are symmetrical about the main diagonal. Each section consists of one row (or column) for each of the m constraints and one column (or row) for each of the n independent variables. The entries in each row (or column) are the first order partials of that constraint with respect to each of the independent variables. Thus the upper right section will be an $m \times n$ array and the lower left section will be an $n \times m$ array. The lower left section is an $n \times n$ array that contains the values for the second order partials of F. The entries in the lower right section are the same values as are in the original Hessian determinant.

Now examine the principal minors of the bordered Hessian determinant. Because of the border, and in particular the section of zeroes, we must begin our examination with a higher principal minor. Specifically we must begin with $|\overline{H}_{m+1}|$, which is the $m + 1$ principal minor of the Hessian section, *plus the border*. Adding the border to $|\overline{H}_{m+1}|$ effectively makes the size of the array $(2m + 1) \times (2m + 1)$. The array for $|\overline{H}_{m+2}|$ will be $(2m + 2) \times (2m + 2)$ and so on to $|\overline{H}_n|$. After examining the principal minors, one of these conditions must hold:

1. When $|\overline{H}_{m+1}|, |\overline{H}_{m+2}|, \ldots, |\overline{H}_n|$ all have the same sign, $(-1)^m$, the critical point is a minimum.
2. When the sign of $|\overline{H}_{m+1}|, |\overline{H}_{m+2}|, \ldots, |\overline{H}_n|$ alternates, beginning with $(-1)^{m+1}$, the critical point is a maximum.
3. When neither of the first two conditions holds, the test fails and we must examine a neighborhood around the critical point to determine whether an extremum exists there.

Example: Optimize $f(x_1, x_2, x_3) = 10x_1x_2x_3$ subject to $g(x_1, x_2, x_3) = 120 - 4x_1 - 2x_2 - 4x_3 = 0$ using a bordered Hessian determinant. Begin by constructing $F(x_1, x_2, x_3, \lambda) = 10x_1x_2x_3 - \lambda(120 - 4x_1 - 2x_2 - 4x_3)$. To satisfy the first order condition, take the first order partials of F, set them equal to zero, and solve:

$$F_{x_1} = 10x_2x_3 + 4\lambda = 0$$
$$F_{x_2} = 10x_1x_3 + 2\lambda = 0$$
$$F_{x_3} = 10x_1x_2 + 4\lambda = 0$$
$$F_\lambda = -120 + 4x_1 + 2x_2 + 4x_3 = 0$$

I will forego the details of the solution, but you should be able to manipulate the equations to get $x_1 = 10$, $x_2 = 20$, $x_3 = 10$, and $\lambda = -500$.

Next take the appropriate partial derivatives and construct the bordered Hessian:

$$|\overline{H}| = \begin{vmatrix} 0 & 4 & 2 & 4 \\ 4 & 0 & 10x_3 & 10x_2 \\ 2 & 10x_3 & 0 & 10x_1 \\ 4 & 10x_2 & 10x_1 & 0 \end{vmatrix} = \begin{vmatrix} 0 & 4 & 2 & 4 \\ 4 & 0 & 100 & 200 \\ 2 & 100 & 0 & 100 \\ 4 & 200 & 100 & 0 \end{vmatrix}$$

Now examine the principal minors of the bordered Hessian determinant. We begin with $|\overline{H}_2|$ because $m = 1$:

$$|\overline{H}_2| = \begin{vmatrix} 0 & 4 & 2 \\ 4 & 0 & 100 \\ 2 & 100 & 0 \end{vmatrix} = 1{,}600, \text{ and}$$

$$|\overline{H}_3| = \begin{vmatrix} 0 & 4 & 2 & 4 \\ 4 & 0 & 100 & 200 \\ 2 & 100 & 0 & 100 \\ 4 & 200 & 100 & 0 \end{vmatrix} = -480{,}000$$

The signs alternate beginning with $(-1)^{1+1} = 1$, which indicates that the critical point, (10, 20, 10, 20,000), is a maximum.

Like most of the concepts presented in this monograph, bordered Hessian determinants are used in regression analysis. Although a computer program will do the calculations, you need to understand the basis of the calculations so you can properly interpret the results or correct any data errors.

ABOUT THE AUTHOR

TIMOTHY M. HAGLE is currently associate professor in the Department of Political Science at the University of Iowa. He received his B.S. in Mathematics, B.A. in Communications (both with teaching certificate), M.A. and Ph.D. in Political Science from Michigan State University, and his J.D. from Thomas M. Cooley Law School. He has taught mathematics to high school and undergraduate students, and he currently teaches the material in this monograph to first-year graduate students. He has published several articles in the area of judicial politics and behavior. His current research interests are in United States Supreme Court decision making and behavior.

Quantitative Applications
in the Social Sciences
A SAGE UNIVERSITY PAPERS SERIES
$9.75 each
Place
Stamp
here
SAGE PUBLICATIONS, INC.
P.O. BOX 5084
THOUSAND OAKS, CALIFORNIA 91359-9924

Quantitative Applications in the Social Sciences

A SAGE UNIVERSITY PAPERS SERIES

63. **Survey Questions** *Converse/Presser*
64. **Latent Class Analysis** *McCutcheon*
65. **Three-Way Scaling and Clustering** *Arabie/Carroll/DeSarbo*
66. **Q Methodology** *McKeown/Thomas*
67. **Analyzing Decision Making** *Louviere*
68. **Rasch Models for Measurement** *Andrich*
69. **Principal Components Analysis** *Dunteman*
70. **Pooled Time Series Analysis** *Sayrs*
71. **Analyzing Complex Survey Data** *Lee/Forthofer/Lorimor*
72. **Interaction Effects in Multiple Regression** *Jaccard/Turrisi/Wan*
73. **Understanding Significance Testing** *Mohr*
74. **Experimental Design and Analysis** *Brown/Melamed*
75. **Metric Scaling** *Weller/Romney*
76. **Longitudinal Research** *Menard*
77. **Expert Systems** *Benfer/Brent/Furbee*
78. **Data Theory and Dimensional Analysis** *Jacoby*
79. **Regression Diagnostics** *Fox*
80. **Computer-Assisted Interviewing** *Saris*
81. **Contextual Analysis** *Iversen*
82. **Summated Rating Scale Construction** *Spector*
83. **Central Tendency and Variability** *Weisberg*
84. **ANOVA: Repeated Measures** *Girden*
85. **Processing Data** *Bourque/Clark*
86. **Logit Modeling** *DeMaris*
87. **Analytic Mapping and Geographic Databases** *Garson/Biggs*
88. **Working With Archival Data** *Elder/Pavalko/Clipp*
89. **Multiple Comparison Procedures** *Toothaker*
90. **Nonparametric Statistics** *Gibbons*
91. **Nonparametric Measures of Association** *Gibbons*
92. **Understanding Regression Assumptions** *Berry*
93. **Regression With Dummy Variables** *Hardy*
94. **Loglinear Models With Latent Variables** *Hagenaars*
95. **Bootstrapping** *Mooney/Duval*
96. **Maximum Likelihood Estimation** *Eliason*
97. **Ordinal Log-Linear Models** *Ishii-Kuntz*
98. **Random Factors in ANOVA** *Jackson/Brashers*
99. **Univariate Tests for Time Series Models** *Cromwell/Labys/Terraza*
100. **Multivariate Tests for Time Series Models** *Cromwell/Hannan/Labys/Terraza*
101. **Interpreting Probability Models:** Logit, Probit, and Other Generalized Linear Models *Liao*
102. **Typologies and Taxonomies** *Bailey*
103. **Data Analysis** *Lewis-Beck*
104. **Multiple Attribute Decision Making** *Yoon/Hwang*
105. **Causal Analysis With Panel Data** *Finkel*
106. **Applied Logistic Regression Analysis** *Menard*
107. **Chaos and Catastrophe Theories** *Brown*
108. **Basic Math for Social Scientists:** Concepts *Hagle*

ISBN 0-8039-5875-7
90000>
9 780803 958753

SAGE PUBLICATIONS
International Educational and Professional Publisher
Thousand Oaks London New Delhi